Liv

Living with the shore

Series editors, Orrin H. Pilkey, Jr. and William J. Neal

The beaches are moving: the drowning of America's shoreline
New edition   Wallace Kaufman and Orrin H. Pilkey, Jr.

From Currituck to Calabash: Living with North Carolina's barrier islands *Second edition*   Orrin H. Pilkey, Jr., et al.

Living with the South Carolina shore
William J. Neal et al.

Living with the East Florida shore
Orrin H. Pilkey, Jr., et al.

Living with the West Florida shore
Larry J. Doyle et al.

Living with the Alabama-Mississippi shore
Wayne F. Canis et al.

Living with the Louisiana shore
Joseph T. Kelley et al.

Living with the Texas shore
Robert A. Morton et al.

Living with Long Island's south shore
Larry McCormick et al.

Living with the New Jersey shore
Karl F. Nordstrom et al.

Living with the California coast
Gary Griggs and Lauret Savoy et al.

Living with the shore of Puget Sound and the Georgia Strait   Thomas A. Terich

Living with the Lake Erie shore
Charles H. Carter et al.

Living with the Coast of Maine
Joseph T. Kelley et al.

Living with the Chesapeake Bay and Virginia's ocean shores

Larry G. Ward   Peter S. Rosen   William J. Neal
Orrin H. Pilkey, Jr.   Orrin H. Pilkey, Sr.   Gary Anderson
Stephen J. Howie

Sponsored by the National Audubon Society ™

Duke University Press   Durham and London   1989

The National Audubon Society and Its Mission

In the late 1800s, forward-thinking people became concerned over the slaughter of plumed birds for the millinery trade. They gathered together in groups to protest, calling themselves Audubon societies after the famous painter and naturalist John James Audubon. In 1905, thirty-five state Audubon groups incorporated as the National Association of Audubon Societies for the Protection of Wild Birds and Animals, since shortened to National Audubon Society. Now, with more than half a million members, five hundred chapters, ten regional offices, a twenty-five million dollar budget, and a staff of two hundred seventy-three, the Audubon Society is a powerful force for conservation, research, education, and action.

The Society's headquarters are in New York City; the legislative branch works out of an office on Capitol Hill in Washington, D.C. Ecology camps, environmental education centers, research stations, and eighty sanctuaries are strategically located around the country. The Society publishes a prize-winning magazine, *Audubon*, an ornithological journal, *American Birds*, a newspaper of environmental issues and Society activities, *Audubon Action*, and a newsletter as part of the youth education program, *Audubon Adventures*.

The Society's mission is expressed by the Audubon Cause: to conserve plants and animals and their habitats, to further the wise use of land and water, to promote rational energy strategies, to protect life from pollution, and to seek solutions to global environmental problems.

National Audubon Society 950 Third Avenue New York, New York 10022

Living with the Shore Series

Publication of the various volumes in the Living with the Shore series has been greatly assisted by the following individuals and organizations: the American Conservation Association, an anonymous Texas foundation, the Charleston Natural History Society, the Coastal Zone Management Agency (NOAA), the Geraldine R. Dodge Foundation, the William H. Donner Foundation, Inc., the Federal Emergency Management Agency, the George Gund Foundation, the Mobil Oil Corporation, Elizabeth O'Connor, the Sapelo Island Research Foundation, the Sea Grant programs in New Jersey, North Carolina, Florida, Mississippi/Alabama, and New York, The Fund for New Jersey, M. Harvey Weil, and Patrick H. Welder, Jr. The Living with the Shore series is a product of the Duke University Program for the Study of Developed Shorelines, which is funded by the Donner Foundation.

© 1989 Duke University Press, all rights reserved
Printed in the United States of America on acid-free paper ∞
Library of Congress Cataloging-in-Publication Data
Living with Chesapeake Bay and Virginia's ocean shores / [edited] by
Larry G. Ward . . . [et al.].   p. cm. — (Living with the shore)   Bibliography: p.
ISBN 0-8223-0868-1. ISBN 0-8223-0889-4 (pbk.)
1. Shore protection—Maryland.   2. Shore protection—Virginia.
3. Coastal zone management—Maryland.   4. Coastal zone management—
Virginia.   5. Coasts—Maryland.   6. Coasts—Virginia.   7. House
construction—Maryland.   8. House construction—Virginia.   I. Ward,
Larry G.   II. Series.
TC224.M3L58   1989   333.91'716'09752—dc19   88-21738 CIP

Contents

List of figures and tables viii
Foreword xiii

1. The dynamic coast 1

Origin of Chesapeake Bay 6
Sea-level rise: a cause of shore erosion 7
   The accelerating rise in sea level 8
Tides 10
Waves: the agent of shore erosion 11
Coastal flooding 12
Storms: the time for shore erosion 16
   Nor'easters 17
   Tropical storms (including hurricanes) 18
Impacts of storms on shorelines 20
A story of shoreline development 21
The moral of the story 24

2. Coastal environments 25

Shore types within Chesapeake Bay 25
   Bluffs: erosional scarps 25
   Marsh shorelines: a natural defense 26
   Estuarine beaches: the wave buffer 29
   Barrier beaches and spits 31
   Dunes: stored sand 31
   Lowlands 32
   Man-modified shores: artificial stabilization 32
The open-ocean coast 33
   The origin of barrier islands 33
   Barrier islands: islands on the move 34
   Accelerating sea-level rise 35
   Barrier-island migration 35
   Ocean-facing beaches: the dynamic equilibrium 39
   How does the beach respond to a storm? 39
   How does the beach rebuild after a storm? 40
   The long-range future of beach development 42
   Erosion on the back sides of our islands 42

3. Shoreline engineering: stabilizing the unstable 43

Solutions to coastal erosion 47
   Do nothing 48
   Construction setback 48
   Move back 48
   Natural solutions 50

vi   Contents

    Beach nourishment: a "soft" solution   52
    Hard solutions   53
A philosophy of shoreline conservation   71

4. Selecting a site along the shores of Chesapeake Bay   73

Shoreline descriptions and site-analysis maps   74
    Using the site-analysis maps   76
Eastern Shore, Virginia: Northampton and Accomack counties   78
Lower Eastern Shore, Maryland: Somerset, Wicomico, and lower Dorchester counties   79
Middle Eastern Shore, Maryland: upper Dorchester, Talbot, and Queen Annes counties   83
Upper Eastern Shore, Maryland: upper Queen Annes, Kent, and Cecil counties   87
Upper Western Shore, Maryland: Harford, Baltimore, and upper Anne Arundel counties   93
Lower Western Shore, Maryland: lower Anne Arundel, Calvert, and Saint Marys counties   97
Lower Western Shore, Virginia: Northumberland, Lancaster, Middlesex, Mathews, Gloucester, and York counties, Hampton Roads, and Norfolk   103

5. The open-ocean shoreline   148

Delmarva's barrier islands   148
Virginia Beach: facing the sea   150
Sandbridge   153
Coastal development   154

6. Coastal land use and the law   155

Federal programs   156
    National Flood Insurance Program   156
    Some flood-insurance facts   158
    Hurricane evacuation   159
    Coastal Barrier Resources Act of 1982   159
    Navigable waterways   160
    Federal Coastal Zone Management Act   160
Local and state laws   160
    Critical Area Law (Maryland)   161
    Pending land-use regulation (Virginia)   162
    Zoning and subdivision (all areas)   162
    Building permits (all areas)   163
    Sedimentation control (all areas)   163
    Water supply and sewerage (all areas)   163
    General wetlands protection   163
    Maryland Wetlands Act   164
    Virginia Wetlands Act   164

7. Building or buying a house near the shore   166

Coastal realty versus coastal reality   166
The structure: concept of balanced risk   167
Coastal forces: design requirements   168

 Wind 168
 Storm surge and flooding 169
 Waves 172
 Battering by debris 172
 Barometric-pressure change 172
House selection 172
Keeping dry: pole or stilt houses 173
An existing house: what to look for 178
 Geographic location 178
 Quality of construction 178
 Improvement possibilities 184
Mobile homes: limiting their mobility 186
High-rise buildings: the urban shore 187
Modular-unit construction: prefabricating the urban shore 189
Living with nature: prudence pays 191

Appendix A Hurricanes 192

Ranking hurricanes: how bad is bad? 192
Hurricane checklist 193

Appendix B A guide to federal, state, and local agencies involved in coastal development 198

Appendix C Useful references 204

Index 228

Figures and tables

Figures

1.1 Location map of Chesapeake Bay and surrounding area 2
1.2 Typical cross sections of Chesapeake Bay for the upper, middle, and lower reaches 3
1.3 Historic shoreline changes for Poplar, Tilghman, Sharp, and James islands 5
1.4 Sea-level rise for the past 30,000 years 6
1.5 Sea-level rise for the past 8,000 years for the Delaware coast 7
1.6 Tide gauge records for Baltimore, Annapolis, and Solomons, Maryland, and Hampton Roads, Virginia 9
1.7 Subsidence (sinking of the land) rates in the Chesapeake Bay region 9
1.8 Relationship between horizontal shoreline movement (erosion) and vertical sea-level rise 10
1.9 Tidal ranges for Chesapeake Bay 11
1.10 Relationship between the relative position of the sun, moon, and earth and the phase of the astronomic tide (spring and neap) 12
1.11 Relationship between wave approach and longshore drift of sediment 13
1.12 Groin field along Hampton, Virginia 14
1.13 Multiple nearshore bars near Silver Beach, Virginia 14
1.14 Relationship between observed tidal height, predicted tidal height, and a storm surge 15
1.15 Predicted 100-year storm surge heights in Maryland's Chesapeake Bay 15
1.16 Storm waves breaking on shoreline at Yorktown, Virginia 17
1.17 Probability (percentage) that a hurricane will occur in any one year for various locations on the U.S. East Coast and Gulf Coast 19
1.18 Twentieth-century hurricane tracks in the vicinity of the Chesapeake Bay and Virginia barrier island coast 20
1.19 Relationship between the approach of a storm and flooding in Chesapeake Bay 22
2.1 High bluffs located at Calvert Cliffs, Maryland 26
2.2 Infiltration of groundwater can lead to bluff failure 27
2.3 Large marsh near Crisfield, Maryland 27
2.4 Blackwater Wildlife Refuge, Maryland 28
2.5 Marsh fronted by beach south of Gwynn Island, Virginia 29
2.6 Small fringing marsh in Accomack County, Virginia 29
2.7 Estuarine beach located along the Choptank River, Maryland 30

Figures and tables ix

2.8 Dune fields located at Deal Island, Maryland, and Russell Island, Virginia 32
2.9 Extensive lowlands located near the mouth of the Choptank River 33
2.10 Major environments found along Delmarva barrier islands and Smith Island, Virginia 34
2.11 Stages of barrier development during the Holocene sea-level rise 36
2.12 Landward movement of barrier islands across the continental shelf as sea level rose over approximately the last 15,000 years 37
2.13 Lagoonal system (Metomkin Bay) located landward of Metomkin Island, Virginia 38
2.14 Typical southern Delmarva tidal inlet (Wachapreague Inlet and associated shoals) 39
2.15 Factors affecting the dynamic equilibrium of a coastal system 40
2.16 Changes in a beach following a storm 41
3.1 The effects of uncoordinated shore protection structures along a coastal reach 45
3.2 Coastal structures not properly tied in to the fastland often suffer end-around effects 46
3.3 Marsh grasses planted in a low-wave energy cove along the Choptank River are stabilizing this shoreline 50
3.4 Extensive dunes and sand fences located on Assateague Island, Virginia 51
3.5 Sand trucking and sand pumping at Virginia Beach 52
3.6 Stone and timber groins 54
3.7 Various groins and a groin field in the Chesapeake Bay area 55
3.8 Jetties at Ocean City and along the Choptank River, Maryland 56
3.9 Common types of bulkheads found around Chesapeake Bay 58
3.10 Various bulkheads in the Chesapeake Bay area 60
3.11 Various types of revetments found around Chesapeake Bay 62
3.12 Stone revetments in the Chesapeake Bay area 63
3.13 Influence of waves and overtopping on bulkheads 64
3.14 Failed bulkheads in Chesapeake Bay 65
3.15 Influence of graded stone or filter cloths on revetments 66
3.16 Grading a steep bank reduces slumping and erosion 67
3.17 Overhead view of a breakwater 68
3.18 Cross sections of common types of breakwaters 69
3.19 Breakwaters located at Colonial Beach, Virginia 70
3.20 Perched beach construction: cross-sectional view and plan view 71
3.21 Sand-filled bags being used as shore protection in Chesapeake Bay 72
4.1 Map showing boundaries of site-analysis maps 77
4.2 Site analysis: Northampton County, Virginia 80
4.3 Site analysis: Accomack County, Virginia 84
4.4 Eroding bluff in Northampton County, Virginia 87
4.5 Site analysis: lower Somerset County, Maryland 88
4.6 Site analysis: middle Somerset County, Maryland 89

x Figures and tables

4.7 Site analysis: upper Somerset and Wicomico counties, Maryland 91
4.8 Site analysis: lower Dorchester County, Maryland 92
4.9 Site analysis: Bloodsworth, South Marsh, Smith, and Tangier islands 94
4.10 Site analysis: Dorchester County, Maryland (Taylors and Hooper islands) 98
4.11 Site analysis: Dorchester County, Maryland (Little Choptank and Choptank rivers) 101
4.12 Site analysis: Talbot County, Maryland (Choptank River) 102
4.13 Site analysis: Talbot County, Maryland (Tilghman Island and Miles River) 104
4.14 Cambridge, Maryland 105
4.15 Site analysis: Talbot and Queen Annes counties, Maryland (Eastern and Prospect bays) 106
4.16 Site analysis: Kent Island, Maryland 108
4.17 Site analysis: Chester River, Maryland 110
4.18 Site analysis: middle and lower Kent County, Maryland 112
4.19 Site analysis: upper Kent and Cecil counties, Maryland 114
4.20 Site analysis: Cecil County, Maryland 116
4.21 Site analysis: upper Harford County, Maryland 117
4.22 Site analysis: lower Harford and upper Baltimore counties, Maryland 118
4.23 Site analysis: Baltimore County, Maryland 119
4.24 Site analysis: Baltimore Harbor, Maryland 121
4.25 Site analysis: upper Anne Arundel County, Maryland 122
4.26 Site analysis: middle Anne Arundel County, Maryland 123
4.27 Site analysis: lower Anne Arundel County, Maryland 125
4.28 Site analysis: lower Anne Arundel and upper Calvert counties, Maryland 126
4.29 Site analysis: middle and lower Calvert County, Maryland 128
4.30 Site analysis: Saint Marys County, Maryland 130
4.31 Site analysis: upper Northumberland County, Virginia 132
4.32 Site analysis: lower Northumberland and Lancaster counties, Virginia 134
4.33 Site analysis: lower Lancaster County, Virginia 136
4.34 Site analysis: Middlesex, Gloucester, and Mathews counties, Virginia (Rappahannock to the Piankatank River) 137
4.35 Site analysis: Mathews County, Virginia 140
4.36 Site analysis: Mobjack Bay, Virginia 141
4.37 Site analysis: York County and Hampton, Virginia 142
4.38 Site analysis: Hampton, Newport News, and Norfolk, Virginia 144

4.39 Site analysis: Willoughby Spit to Cape Henry, Virginia   145
5.1   Index map of the Delmarva barrier islands and Virginia Beach   149
5.2   Virginia Beach's shoreline is one of the most heavily developed U.S. barrier shorelines   151
5.3   Dumping of sand on Virginia Beach from trucks helps maintain this resort beach   153
5.4   Sandbridge, Virginia   154
7.1   Forces to be reckoned with at the shore   170
7.2   Modes of failure and how to deal with them   171
7.3   Shallow and deep supports for poles and posts   174
7.4   Pole house construction at Sandbridge, Virginia, and poles in place prior to construction at Virginia Beach   175
7.5   Framing system for an elevated house   176
7.6   Tying floors to poles   177
7.7   Foundation anchorage   179
7.8   Stud-to-floor, plate-to-floor framing methods   180
7.9   Roof-to-wall connectors   180
7.10 Where to strengthen a house   182
7.11 Reinforced tie beam (bond beam) for concrete block walls   183
7.12 Some rules in selecting or designing a house   185
7.13 Tiedowns for mobile homes   188
7.14 Hardware for mobile home tiedowns   189

Tables

1.1   Predicted storm surges from 100-year storms in Chesapeake Bay   16
1.2   Recent Chesapeake Bay storm surges   17
3.1   Common erosion-abatement techniques   49
4.1   Classification of erosion rates used on site-analysis maps for Maryland and Virginia's Chesapeake Bay shoreline   76
4.2   Shoreline types and erosion rates: Northampton County, Virginia   82
4.3   Shoreline types and erosion rates: Accomack County, Virginia   86
4.4   Erosion rates: lower and middle Somerset County, Maryland   90
4.5   Erosion rates: upper Somerset and Wicomico counties and lower Dorchester County, Maryland   93
4.6   Erosion rates: Bloodsworth, South Marsh, Smith, and Tangier islands   96
4.7   Erosion rates: Dorchester County, Maryland (Taylors and Hooper islands)   100
4.8   Erosion rates: Dorchester County, Maryland (Little Choptank and Choptank rivers)   100
4.9   Erosion rates: Talbot County, Maryland (Choptank River)   103
4.10 Erosion rates: Talbot County, Maryland (Tilghman Island and Miles River)   105

xii Figures and tables

4.11 Erosion rates: Talbot and Queen Annes counties, Maryland (Eastern and Prospect bays)   107
4.12 Erosion rates: Kent Island, Maryland   109
4.13 Erosion rates: Chester River, Maryland   111
4.14 Erosion rates: lower and middle Kent County, Maryland   113
4.15 Erosion rates: upper Kent and Cecil counties, Maryland   115
4.16 Erosion rates: Cecil County, Maryland   120
4.17 Erosion rates: upper Harford County, Maryland   120
4.18 Erosion rates: lower Harford and upper Baltimore counties, Maryland   120
4.19 Erosion rates: Baltimore County, Maryland   120
4.20 Erosion rates: Baltimore Harbor, Maryland   124
4.21 Erosion rates: upper Anne Arundel County, Maryland   124
4.22 Erosion rates: middle Anne Arundel County, Maryland   124
4.23 Erosion rates: lower Anne Arundel County, Maryland   127
4.24 Erosion rates: lower Anne Arundel and upper Calvert counties, Maryland   127
4.25 Erosion rates: middle and lower Calvert County, Maryland   128
4.26 Erosion rates: Saint Marys County, Maryland   131
4.27 Shoreline types and erosion rates: upper Northumberland County, Virginia   133
4.28 Shoreline types and erosion rates: lower Northumberland and Lancaster counties, Virginia   135
4.29 Shoreline types and erosion rates: lower Lancaster County, Virginia   138
4.30 Shoreline types and erosion rates: Middlesex, Gloucester, and Mathews counties, Virginia (Rappahannock to the Piankatank River)   138
4.31 Shoreline types and erosion rates: Mathews County, Virginia   143
4.32 Shoreline types and erosion rates: Mobjack Bay, Virginia   143
4.33 Shoreline types and erosion rates: York County and Hampton, Virginia   146
4.34 Shoreline types and erosion rates: Hampton, Newport News, and Norfolk, Virginia   146
4.35 Shoreline types and erosion rates: Willoughby Spit to Cape Henry, Virginia   147
7.1 Tiedown anchorage requirements   190
A.1 Saffir-Simpson Hurricane Scale   193

Foreword

Shoreline erosion and flooding are problems that coastal dwellers face nearly everywhere. Along Chesapeake Bay and Virginia's ocean shores the story is no different. In fact, rates of shoreline erosion there rival those found anywhere along the East Coast of the United States. During the past several decades a great deal has been learned about nearshore processes, coastal engineering, and the environmental consequences of developing coastal areas for private homes, condominiums, harbor facilities, and industries. With this information in hand, wise and prudent development of our shoreline is possible.

However, many past mistakes continue to be made. That is why this book was written: to help citizens of Maryland and Virginia understand how development affects their beautiful coastline.

Intended for coastal dwellers, coastal developers, and those charged with coastal regulation, the book presents relevant information concerning Chesapeake Bay and its nearby Atlantic Ocean shores. Among the topics discussed are the origin of the Bay and the ocean coast, sea-level rise, storm effects, coastal engineering, construction near the shore, erosion rates, flooding problems, and applicable laws and regulations. In addition, we point out sources of further important information for those who need it.

The authors are a mixed group. Larry Ward is a research scientist at the University of Maryland. Peter Rosen is a professor of geology at Northeastern University. William Neal is a professor of geology at Grand Valley State University, Michigan. Orrin Pilkey, Jr., is a James B. Duke professor of geology at Duke University. Orrin Pilkey, Sr., is a retired civil engineer who lives in Virginia. Gary Anderson is a coastal consultant with Espey, Huston, and Associates. And Stephen Howie is with the U.S. Environmental Protection Agency.

With publication of *Living with the Chesapeake Bay and Virginia's Ocean Shores*, the Living with the Shore series now boasts 14 books, all of which are listed on the inside cover of this volume. The classic *The Beaches Are Moving: The Drowning of America's Shoreline*, by Wallace Kaufman and Orrin Pilkey, Jr., covers the basic issues dealt with in this and the other state-by-state books.

The overall coastal book project is an outgrowth of initial support from the National Oceanic and Atmospheric Administration (NOAA) through the Office of Coastal Zone Management. The initial project was administered through the North Carolina Sea Grant Program. More recently it has been generously supported by the Federal Emergency Management Agency (FEMA). Without FEMA support the series would not have proceeded this far. However, the conclusions presented herein are those of the authors and do not necessarily represent those of the supporting agencies.

We owe a debt of gratitude to many individuals for support, ideas, encouragement, and information. Doris Schroeder has helped us in many ways as Jill-of-all-trades over a span of more than a decade and more than a dozen books. The Duke University Press staff compiled the index for this volume. The original idea for a coastal book (*How to Live with an Island* [1972]) was that of Pete Chenery, then director of the North Carolina Science and Technology Center. Richard Foster of the Federal Coastal Zone Management Agency supported the project at a critical juncture. Because of his lifelong commitment to land conservation, Richard Pough, former head of The Nature Conservancy, has been a mainstay in our fund-raising efforts. Myrna Jackson of the Duke University Development Office has been most helpful in our search for financial support.

Jane Bullock of FEMA has been a constant source of encouragement in helping us chart a course through the federal government's shifting channels. Richard Krimm, Peter Gibson, Dennis Carroll, Jim Collins, Jet Battley, Melita Rodeck, Chris Makris, and many others have opened doors, provided maps, charts, and publications, and generally assisted us through the Washington maze.

As with any effort of this sort, we were helped by many people who live and work by the shore—so many we cannot begin to list them all. We are grateful for their cooperation, insight, and concern.

Last but not least, we extend our thanks to Lynne Claflin, Virginia Henderson, and Tonya Clayton for doing a lot of the drafting and manuscript/figure preparation. In addition, Jane Gilliard and Anna Ruth McGinn typed the manuscript.

We dedicate this work to all those who helped us and to all who enjoy the beautiful shoreline of Chesapeake Bay.

Orrin H. Pilkey, Jr.
William J. Neal
Series Editors

1 The dynamic coast

The placid nature of Chesapeake Bay has been heralded since Captain John Smith first arrived in 1607, but its calmness, beauty, and serenity can be deceiving. The Bay's shoreline is strikingly dynamic. If Captain Smith could visit Chesapeake Bay once again, he probably would recognize very little—not because of the development, but because much of the shoreline he mapped is no longer there! Many of today's beaches are 200 to 2,000 feet landward of the shore that he saw. The settlements he knew on the shore are under water today. Archaeologists do much of their digging to find artifacts of the coastal settlement of Jamestown in a salt marsh below the high-tide line.

Although the Bay is one of the largest estuaries in the world, it is actually a very shallow body of water. As sea level has risen, it has flooded broad river valleys to create a highly dissected shoreline (fig. 1.1). In the 180-mile length of the Bay, the total amount of shoreline is over 7,000 miles (reference 4, appendix C). The average water depth in the Bay is on the order of 20 to 25 feet. Deeper water is found only in the original river channels that now form the shipping lanes down the center of the Bay. Hence, the Bay in cross section is a shallow pan creased by a narrow channel (fig. 1.2). Professor M. G. Wolman of Johns Hopkins University suggests that this form can be visualized this way: If the width of the bay is the width of this page, the depth of the bay is about one third the thickness of the paper (reference 27, appendix C).

Current development at the shore often overlooks the Bay's formation and the dynamics of its operation, including erosion and flooding. The shores that John Smith knew are lost without record, but more recent shores that are now missing should not be forgotten. Older Eastern Shore watermen know the shoal off the mouth of the Choptank River was once Sharps Island, but few know the stories of the island's days as a resort. At the turn of this century the island was a favorite spot for hunting, and an artesian well supplied water for the resort's hotel. By 1910 the hotel stood abandoned on only 53 acres of island, and the well site was in the waters of Chesapeake Bay. In a 1914 study J. F. Hunter predicted that the island would be gone by 1951 (reference 105, appendix C). His prediction was fairly accurate as the 1965 charts show no remnants of the island! The 438-acre island had disappeared in a little over a century, along with its forest, its buildings, its well, and most of its memories.

Wouldn't it be convenient if the loss of Sharps Island were an isolated case, a fluke of nature due to some unusual convergence of currents or waves? In fact, it is not, and Sharps Island is not alone in the pattern of its shrinking. Poplar Island is now only a set

2 The dynamic coast

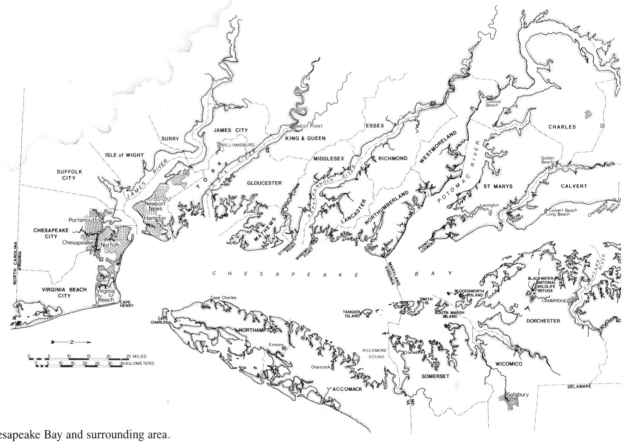

1.1 Location map of Chesapeake Bay and surrounding area.

The dynamic coast 3

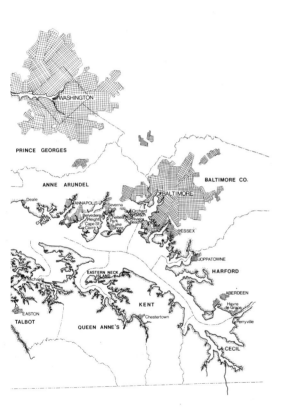

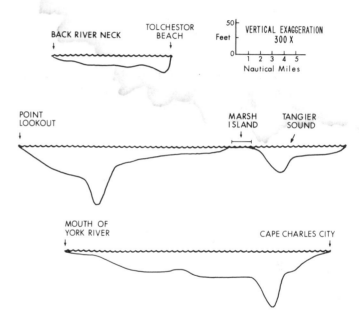

1.2 Typical cross sections of Chesapeake Bay for the upper (Back River Neck–Tolchestor Beach), middle (Point Lookout–Tangier Sound), and lower (York River–Cape Charles City) reaches. (Note the vertical scale is 300 times smaller than the horizontal, causing the depths to appear much larger.)

of islets, the dissected remains of a much larger island shown on an 1847 map (fig. 1.3). This island's shore eroded at a rate of nine feet per year over a 116-year period of record. Tilghman Island consisted of 2,015 acres in 1848, but had reduced to 1,686 acres by 1901. Today the south end of the island is nearly separated from the main body by a new strait. James Island consisted of 976 acres in 1848, 555 acres in 1901, and 490 acres in 1910. And so the story goes for both islands and mainland shores.

The history of Chesapeake Bay is one of shoreline retreat because of a rise in sea level. Anyone living on the shoreline today should be aware of this history. Shoreline erosion is often ignored. You can watch a shore for years, or sometimes decades, and not notice changes until your house is threatened by the retreating shore. As they have for over 10,000 years, water levels are steadily rising. The shore retreats when large storms come through and waves erode the shoreline. Storms also cause significant flooding. Such storm events are often considered disasters, but are actually part of the continuous evolution of Chesapeake Bay.

Living in coastal areas around the Bay involves certain risks, including those of erosion and flooding. The intent of this book is to help the property owner understand the risks and recognize that actions can be taken to minimize them.

Although many of the forces acting on the Chesapeake Bay shoreline are similar to those influencing open ocean shorelines, the relative roles of the various forces differ. The Bay beach system is *not* a scaled-down ocean beach. Coastal property owners on Chesapeake Bay are at an advantage over oceanfront property owners. Protecting the Bay shoreline with stabilization structures is often feasible, and such structures may last a reasonable amount of time. However, stabilization structures are not a panacea; they can as easily aggravate the beach erosion problem as solve it. Ironically, many homeowners build at the shore for the aesthetics of the beach and then build structures that contribute to beach erosion.

Understanding the natural processes that are at work along the coast will enhance your enjoyment of the environment, help you locate and protect your property, and sometimes guard your well-being with respect to natural hazards. The two hazards that pose the greatest threat to property are erosion and flooding. The wealth of free expert advice available from both the state and federal governments is crucial for understanding an individual shore site in this region. To evaluate the likelihood of coastal hazards and associated risks for a particular piece of property, one must understand the natural setting. Let's begin with the origin of the Bay.

1.3 (A) Historic shoreline changes for Poplar, Tilghman, Sharp, and James islands. Modified after Shepard and Wanless, 1971 (reference 32, appendix C). (B) Poplar Island in 1981. (C) Tilghman Island–Blackwalnut Point in 1981.

The dynamic coast 5

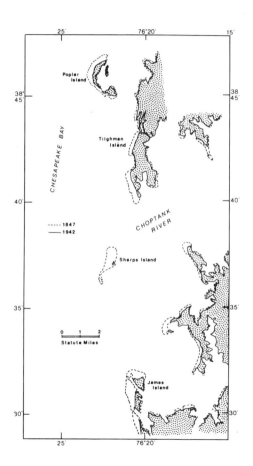

6 The dynamic coast

Origin of Chesapeake Bay

The present Chesapeake Bay was carved by the Susquehanna River and its tributaries during the last great Ice Age as the large glaciers advanced across Canada and Europe into the mid-latitudes. During this period, the glaciers grew at the expense of the ocean, tying up great volumes of water in ice and causing sea level to drop on the order of 150 to 300 feet approximately 15,000 to 20,000 years ago (fig. 1.4). At this time the Susquehanna River, which was being fed by glaciers, flowed across much of the present continental shelf and eroded a great valley into the coastal plain. In places the mighty Susquehanna, helped by its tributaries including the present-day Potomac, Rappahannock, and James, scoured down more than 150 feet. As the glaciers melted, water returned to the oceans and sea level rose. Most evidence indicates that the rising sea invaded the Susquehanna River valley sometime within the last 7,000 to 10,000 years and formed Chesapeake Bay. Sea level probably approached within 10 feet of its present height approximately 2,000 to 3,000 years ago (fig. 1.5). Since that time, changes in the shoreline around the Bay have resulted largely from coastal erosion.

This latest downcutting of the Susquehanna River during low stands of sea level and subsequent flooding as sea level rose is only one of several such episodes that have occurred during the Quaternary (geologic period covering the last 2 million years). At least four major advances of the ice fields have occurred with concurrent lowering of sea level and downcutting of river valleys. Each

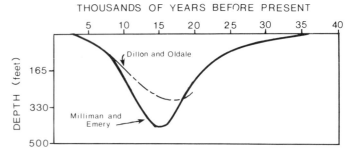

1.4 Sea-level rise for the past 30,000 years. Modified after Dillon and Oldale, 1978 (reference 11, appendix C). Depth refers to changes in the level of the ocean compared to the present mean water level. The Dillon and Oldale curve is based on radio-carbon dated samples from the East Coast of the United States. This curve indicates the ocean levels dropped a maximum of approximately 300 feet.

The dynamic coast   7

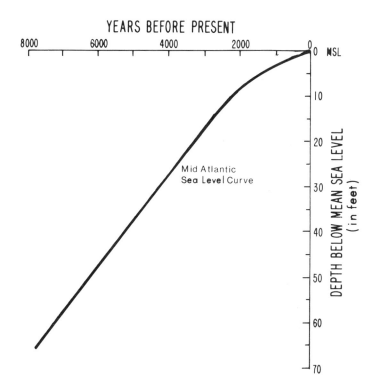

1.5   Sea-level rise for the past 8,000 years for the Delaware coast. Modified after Belknap and Kraft, 1977 (reference 9, appendix C).

time, as the ice sheets melted and the oceans advanced, the river valleys flooded, forming a type of estuary referred to as a *drowned river valley*. Drowned river valleys are excellent sediment traps, capturing almost all the materials eroded from uplands and carried to the estuary. Consequently, the estuary fills in rapidly (geologically speaking), perhaps within 10,000 years. Thus drowned river valleys such as Chesapeake Bay are ephemeral features. Former valleys formed by the Susquehanna River system during previous glaciations have filled in. In some instances, higher stands of sea level have added marine deposits, all but burying the old valleys. Trying to determine the path of the ancestral Susquehanna is a problem that has intrigued geologists for quite some time.

Sea-level rise: a cause of shore erosion

As a result of global warming following the last major glaciation (Wisconsin), most of the ice sheets that covered vast areas of the continents melted. This meltwater eventually was added to the oceans, resulting in a rise in sea level of about 150 to 300 feet in the past 15,000 years (fig. 1.4). The present remainders of these great glaciers are the polar ice caps, which continue to melt. The polar ice caps still retain water sufficient, if melted, to raise sea level approximately another 180 feet.

The formation of the Bay and its shores can be attributed to this ongoing sea-level rise, which flooded ancient river valleys as water levels moved landward. However, if you were standing on the shore after a major storm looking at slumped bluffs or an eroded beach, there is little likelihood that you'd blame rising sea level for your woes!

Most shoreline changes involve sediment moving, either seaward by erosion or downward by slumping. While the rise in sea level does not move material directly, its role is to control the action of other processes. In the presence of a gradually rising sea level, storm surges continually get higher and storm waves continually encroach farther inland. This brings land areas previously safe from shore processes into the coastal zone. So though sea-level rise is a passive factor, it remains very much the controlling factor over the short-term processes (e.g. waves, longshore currents, flooding) responsible for movement of sand and erosion of the shore.

All of this discussion is of little use in evaluating a coastal site unless one is concerned with what sea level will do in a few decades. Tide gauges, used to predict future tides from past records, have provided continuous, accurate records of water levels throughout the Bay for periods up to nearly 100 years (fig. 1.6). The averages of these records show that the water level is currently rising at rates of about one foot per century. If the rise in global sea level were the only change occurring, all of the tide gauges in the Bay should show the same average trend. They do not. The variations are because not only are water levels rising from melting polar ice caps, but the land is sinking at different rates in different places (fig. 1.7). Land is subsiding (sinking) more rapidly in the northeastern portions (Chester River area) and southwestern portions (York, Gloucester, and Mathews counties). Parts of the mouth of the Potomac River (Northumberland County) and the Bay mouth (Northampton County) are subsiding at about half the rate of the other areas.

The accelerating rise in sea level

The National Academy of Science and the Environmental Protection Agency (EPA) have warned that all evidence points to increased warming of the earth's surface. The burning of fossil fuels has increased carbon dioxide and associated combustion gases in the atmosphere, which retains heat (*greenhouse effect*). The net result appears to be a slight warming of the earth's climate, leading to thermal expansion of the oceans and additional melting of continental ice. Ultimately this will cause sea level to rise faster. The EPA has projected that atmospheric warming may cause sea level to rise several feet in the next century. This rise is in addition to changes in water level caused by land subsidence. Consequently, the local rate of sea-level rise may be substantially higher.

The safest assumption about future sea-level rise is that it will continue and will accelerate. Keep in mind that a one-foot-per-century vertical rise in sea level is more significant than it sounds. Figure 1.8 shows that the horizontal distance of shoreline retreat (*erosion*) will be much greater than the distance of vertical rise. How much the shoreline shifts is determined by the slope of the land. For the lowlands along much of the Eastern Shore and topographic lows along some of the tributaries, the horizontal shift of the shoreline can exceed 1,000 feet in a century!

The dynamic coast 9

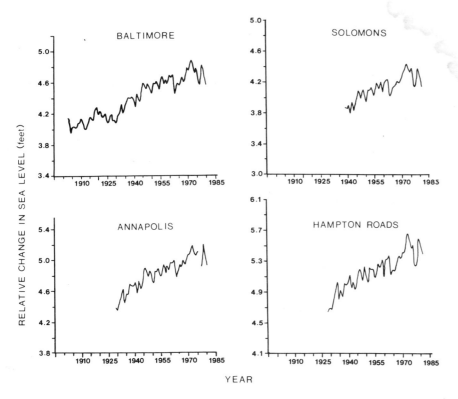
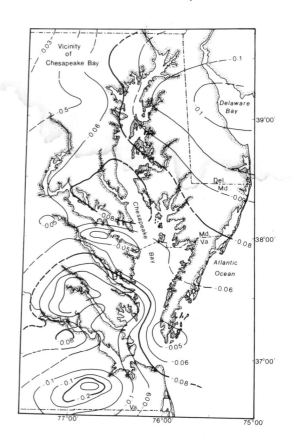

1.6 Tide gauge records for Baltimore, Annapolis, and Solomons, Maryland, and Hampton Roads, Virginia, above. Modified after Hicks et al., 1983 (reference 10, appendix C). **1.7** Subsidence (sinking of the land) rates in the Chesapeake Bay region, right. Modified after Holdahl and Morrison, 1974 (reference 35, appendix C). Contours show rates of subsidence in inches per year.

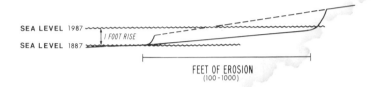

1.8 Relationship between horizontal shoreline movement (erosion) and vertical sea-level rise. Note a small amount of rise in water level can lead to a major amount of shoreline erosion.

If you live high on a western shore cliff, you need not worry about being flooded by such a slow rise in water level. However, a few inches added to the height of the storm waves will allow them to attack the toe of your bluff more frequently. This almost certainly means that the bluff will steepen, and the top of the bluff will continue moving landward.

Tides

Although the tidal range is relatively small throughout Chesapeake Bay, it varies significantly (fig. 1.9). The range at the Bay mouth (Norfolk) is three feet, which is comparable to the adjacent ocean shores. It diminishes to about one foot off Annapolis, and rises to over three feet at the northern end of the Bay. The tidal range is as much as six inches larger on the Eastern Shore than on the Western Shore.

Tidal range not only varies over the length of Chesapeake Bay, it varies from day to day and week to week and over seasons at the same location. Almost everyone who visits the shore of Chesapeake Bay realizes that approximately two high and two low tides occur each day. In addition, most people observe that the tides seem higher and lower (larger tidal range) a few days following full moons and new moons. These higher tides with larger tidal ranges are referred to as *spring tides,* although they occur year-round. Spring tides occur when the earth, moon, and sun are situated in a relatively straight line (fig. 1.10). Although the gravitational attraction of the moon is principally responsible for generating the tides, the sun also has an influence (approximately one-half the moon's). When the sun, moon, and earth are aligned, the sun's and moon's gravitational influence is added together, creating a larger bulge of water and consequently higher and lower tides. Conversely, when the sun and moon form a right angle with the earth (fig. 1.10), they offset each other, decreasing the tidal range. This results in "neap tides." Those who spend a great deal of time on the Bay have most likely also noticed that the spring tides are not always the same. Due to its eccentric orbit around the earth, the moon is closer to the earth approximately every 28 days, causing every other spring tide to be slightly higher. In addition, the earth's orbit around the sun is eccentric, which results in still-higher spring tides when the earth is closest to the sun. The potential for shoreline erosion is at a maximum when the tidal range for a given location is greatest, as waves then attack farther inland.

Although astronomic tides (those tides principally generated by the moon and sun) are extremely predictable, they are only part of the tide-generating force. In Chesapeake Bay winds piling water up or pushing it elsewhere can be as or more effective at creating tides than astronomic forces. On some days there seems to be little or no tide change in parts of Chesapeake Bay. This occurs when wind tides and astronomic tides cancel each other out. More noticeable and potentially more dangerous are those occasions when the wind piles water up and a spring tide occurs, creating extremely high water levels. This will be discussed further when we examine storm surges.

In general, a larger average astronomic tidal range offers greater protection to the shore from the effects of storm surge and flooding. A storm surge must elevate water levels above the high-tide line to have an impact. If all other factors are equal, an area with high tidal range will flood less often than an area with low tidal range. Another factor is the position of the breaking waves, which is controlled by tide levels. In a high tidal-range area, this zone is spread over a larger portion of the beach, which distributes the erosive power of the waves.

Waves: the agent of shore erosion

Waves are generated by wind blowing over the water. The size of a wave is related to the strength and duration of the wind, and the distance of water over which the wind blows (*fetch*). Because fetch is limited in Chesapeake Bay, the waves are much smaller than on ocean shores. Within the Bay, the maximum wave sizes are highly

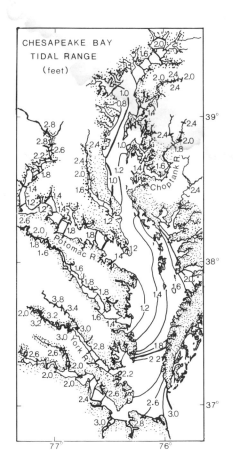

1.9 Tidal ranges for Chesapeake Bay. Modified after Hicks, 1964 (reference 34, appendix C).

12   The dynamic coast

1.10   Relationship between the relative position of the sun, moon, and earth and the phase of the astronomic tide (spring and neap). Blacker areas around the earth represent relative changes in water depths due to tides in the oceans. The changes are greatly exaggerated. Modified from Komar, 1976 (reference 40, appendix C).

variable because of large differences in fetch. If you are evaluating a homesite, the less fetch the better.

Waves approaching the shoreline at an angle create longshore currents that can transport sand along the beach (fig. 1.11). Sand can move alongshore in either direction, but over the long term a "net" transport direction is typically established in response to the largest storm waves. The easiest means of determining net longshore transport direction is to look at a groin (shore perpendicular structure) to see which side has the most sand (figs. 1.11 and 1.12). The sand will accumulate on the side from which the current is coming (*updrift*).

In some areas, storm waves can also transport beach sand offshore to form sandbars (fig. 1.13). Sandbars serve as natural breakwaters to protect the shore. During calm weather, sand stored in the bars may be transported back to the beach.

Many waves formed in Chesapeake Bay have heights of less than one foot. These small waves have little impact on net sediment movement. Most transport of sediment and most shore erosion take place during a few storm events each year when waves in the Bay may reach three- to four-foot heights. These higher waves not only have more energy to move sediment, but often occur on top of storm surges. During storm surges waves break above the normal beach onto dunes, bluffs, or houses.

Coastal flooding

Storm surges are an increase in water level due to the combined effects of wind, low atmospheric pressure, and the shape of the

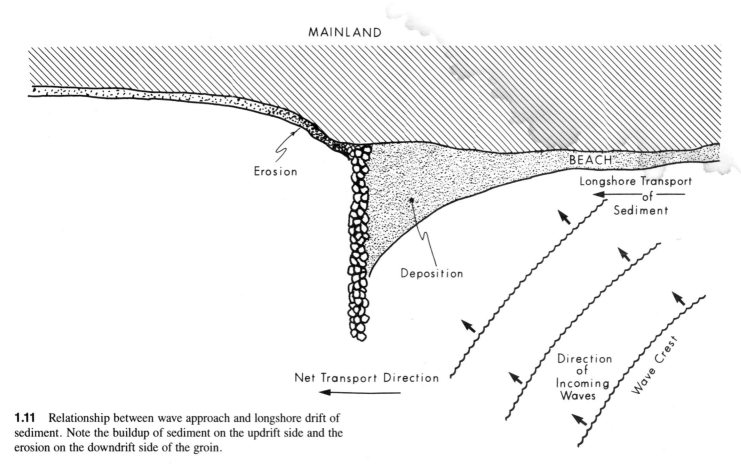

1.11 Relationship between wave approach and longshore drift of sediment. Note the buildup of sediment on the updrift side and the erosion on the downdrift side of the groin.

basin. The surge is mainly due to wind blowing over the water for a long period of time and piling water up against the shore. The resulting water level is the tide level, plus the surge level, plus the height of the waves (fig. 1.14). In a closed basin like the Chesapeake Bay, the effects of a surge can be increased because the water is trapped at one end. The shallow depths throughout most of the Bay make its shores susceptible to flooding during storm surges. Although major storms are responsible for major surges, a strong, steady wind can push a lot of water against the shoreline and generate a surge.

A surge will often grow over a day or more, and then gradually subside. It may last through many tidal cycles with maximum water level occurring at high tide. Of course, the highest floods will occur if the peak surge coincides with the highest spring tides.

For the coastal landowner, the biggest concern is not the cause of coastal flooding, but the extent of the flood (fig. 1.15). Detailed estimates of the height of a *100-year flood* (a flood level with a 1 percent probability of occurring any one year) are available from the Federal Emergency Management Agency (FEMA). FEMA publishes Flood Insurance Rate Maps (FIRMS), which show the areas that most likely will be affected by a 100-year flood. Local, state, or federal government officials as well as local insurance agencies

1.12 Groin field along Hampton, Virginia, above left. Note accumulation of sediment on updrift side of groins (arrows).
1.13 Multiple nearshore bars near Silver Beach, Virginia, below left. Arrow points to bars.

The dynamic coast 15

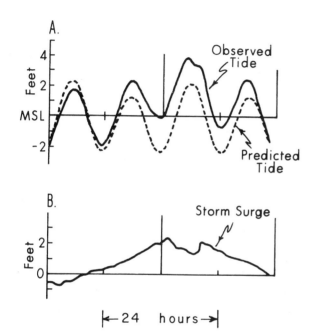

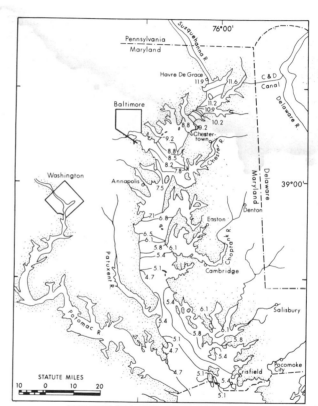

1.14 Relationship between observed tidal height, predicted tidal height, and a storm surge, above. The storm surge is added to the predicted tide (due to astronomic forces), causing the higher actual tidal level. Modified after Pore et al., 1974 (reference 17, appendix C).

1.15 Predicted 100-year storm surge heights in Maryland's Chesapeake Bay, right. Modified after Zabawa and Ostrom, 1982 (reference 89, appendix C).

can direct you to these maps. It is important to recognize that a 100-year flood does not occur once every 100 years. Following a flood of this magnitude, landowners often feel that they're safe for another 99 years. But the so-called 100-year flood could easily occur two years in a row, or even twice in a single year!

The typical storm-surge heights at various locations in Chesapeake Bay are shown in table 1.1. Baltimore often has higher floods than Norfolk, but under the right conditions, any part of the Bay is susceptible to extreme floods.

A cautious estimate of storm-surge height is needed to prepare for worst-case situations. A recently published report by the U.S. Army Corps of Engineers (Baltimore District) calculates maximum surge height during tidal flooding in Chesapeake Bay (reference 14, appendix C). According to the report, the largest tidal flood that is likely to occur under the most severe meteorological and hydrological conditions in Chesapeake Bay is 13 feet above the national geodetic vertical datum. The 13-foot surge excludes wave heights, which under maximum or the worst conditions could reach an additional five feet. Based on these estimates, all shore areas around Chesapeake Bay with an elevation of less than 18 feet would be flooded by such a storm!

Storms: the time for shore erosion

Storm winds are the force behind the most damaging waves. During a storm surge, storm waves not only have more energy, but they break directly against the bluff, dune, or homes behind the beach (fig. 1.16), causing greater amounts of shoreline erosion. If the surge is high, flooding of low-lying areas also occurs. High rainfall and runoff during storms will add to these two problems of flooding and shoreline erosion.

Most storms in the Chesapeake Bay region can be grouped into three general types: summer thunderstorms, nor'easters, and tropical storms (including hurricanes). Summer thunderstorms are locally generated and often produce intense winds that can

Table 1.1 Predicted storm surges from 100-year storms in Chesapeake Bay

Eastern Shore		Western Shore	
Location	Surge (feet)	Location	Surge (feet)
Kiptopeke	5.2	Gloucester Point	5.8
Eastville	4.7	Windmill Point	4.0
Gaskins Point	4.6	Cornfield Harbor	4.6
Guard Shore	6.2	Solomons Island	5.5
Crisfield	5.1	Cove Point	5.2
Chance	5.8	Chesapeake Beach	6.1
Hooper Island	5.5	Annapolis	7.3
Cambridge	6.0	Baltimore	8.1
Matapeake	7.3	Havre De Grace	11.6
Love Point	7.6		
Tolchester	8.7		
Betterton	10.5		

From Boon et al., 1978 (reference 25, appendix C)

associated persistent winds move into a region and may remain for several days. This creates waves and surge effects. If the storm lasts for several days, then the storm surge is likely to occur on top of the normal high tide, possibly more than once, drowning the protective beach and causing severe erosion. If the storm coincides with a spring tide, the effects can be even more spectacular.

In Chesapeake Bay the average occurrence of nor'easters is about three to five times per year (reference 17, appendix C). Fortunately, most of these storms have surge heights of less than three feet above normal water levels, but this flood level is exceeded periodically. To date, the most intense modern nor'easter along the middle-Atlantic coast was the Ash Wednesday storm of 1962 (table 1.2).

1.16 Storm waves breaking on shoreline at Yorktown, Virginia, during a nor'easter in April 1978.

be a hazard to mariners. Though summer thunderstorms can be locally significant, they are usually of short duration and cannot produce the combination of storm surge and large waves that are responsible for widespread erosion.

Nor'easters

Northeast (extratropical) storms are associated with eastward-moving storm fronts that most commonly occur from late fall to early spring. During nor'easters, intense low-pressure centers with

Table 1.2 Recent Chesapeake Bay storm surges

Storm	Tidal elevations (feet above mean sea level)			
	Norfolk	Mid-Bay	Washington	Baltimore
August 1933	8.0	7.3	9.6	8.2
September 1936	7.5	—	3.0	2.3
October 1954 "Hazel"	3.3	4.8	7.3	6.0
August 1955 "Connie"	4.4	4.6	5.2	6.9
August 1955 "Diane"	4.4	4.5	5.6	5.0
April 1956 Nor'easter	6.5	2.8	4.0	3.3
March 1962 Nor'easter	7.4	6.0	—	4.7

From U.S. Army Corps of Engineers, 1974 (reference 4, appendix C)

Tropical storms (including hurricanes)

Storms that originate in tropical to subtropical areas and then move north are known as tropical storms. Severe tropical storms are among the most dangerous types of weather, and many words have evolved to describe them, including "willy-willies" in Australia, "baguios" in the Philippines, and "typhoon" in the western Pacific. The West Indian term, *huracán*, the evil spirit, has evolved to the term used along the East Coast of North America, "hurricane." Officially, a tropical storm becomes a hurricane when wind velocities exceed 74 mph.

Although meteorologists are still seeking answers to the causes and mechanics of tropical storms, they know the basic model of what happens. During the summer the surface waters off the west coast of Africa heat up to at least 79°F. Evaporation produces a layer of warm, moist air over the ocean. This moist air is trapped by warm air coming from the African continent, but some is drawn upward. As the moist air rises, it cools and condenses, releasing heat, which in turn warms the surrounding air and causes it to rise. As a result of the increasing mass of rising air, a low-pressure area forms (a tropical depression), and warm easterly winds rush in to replace the rising air. The effect of the earth's rotation (*Coriolis force*) deflects the air flow towards the right in the Northern Hemisphere, and the counterclockwise rotating air mass begins to take on the familiar shape of a hurricane. Air forced to the middle of the spiral can move upward only, producing a chimneylike column of rising air—the "eye" of the storm.

So, the storm is a type of heat engine, with rising moist air cooling and condensing, releasing heat to cause more air to rise which allows more air to rush in over the sea, an endless source of moisture. Heavy rainfall characterizes the edge of the cloud mass. The strongest winds may exceed 200 mph. The maximum winds of the largest storms to hit coastal areas are generally unknown because wind-measuring instruments are blown away!

Once formed, the hurricane mass begins to track into higher latitudes and may continue to grow in size and strength. The velocity of this tracking movement can vary from nearly stationary to greater than 60 mph. Consider that the diameter of a hurricane ranges from 60 to 600 miles and gale-force winds may extend over most of this area. The total amount of energy released over the thousands of square miles covered by the storm is almost beyond comprehension. No ship or seawall, cottage, condominium, or other static structure is immune from the impact of such forces.

Richard Frank, an administrator with the National Oceanic and Atmospheric Administration, has described the magnitude of the hurricane peril (reference 21, appendix C). Since 1900, over 130 hurricanes have crossed the Atlantic and Gulf coasts. Fifty-three of these have been classified by the National Weather Service as "major" hurricanes—that is, hurricanes with peak winds in excess of 110 mph and storm surges greater than eight feet. Though fewer hurricanes hit the Chesapeake Bay than the Outer Banks of North Carolina (5 to 11 percent chance in a given year [see fig. 1.17]), Virginia Beach and the lower Bay have a 2 percent probability of being affected.

The effects of these storms can be staggering. The two deadliest United States hurricanes in this century killed over 6,000 people in Galveston, Texas, in 1900; and 1,800 people at Lake Okeechobee, Florida, in 1928. One of the costliest hurricane landfalls in the United States was Hurricane Camille, which caused $1.4 billion in damages in 1969 in Mississippi and Louisiana.

More Americans are at risk from a major hurricane today than at the turn of the century when these earlier disasters occurred. The coastal population is increasing at a rate of more than three times the national average. This is due not to the high fertility of United States beach lovers, but rather to an influx of newcomers to coastal areas. A large percentage of these new coastal dwellers have no significant experience with hurricanes. In the last twenty years, only one hurricane with a substantial death toll has hit the East Coast of the United States, and only two have hit the Gulf Coast. Most of the population has little appreciation of the potential destructive power of these storms. The long period between major hurricanes has lulled society into a false sense of complacency.

While Virginia Beach and the ocean-facing barrier islands to the north are fully exposed to the impacts of a hurricane, Chesapeake

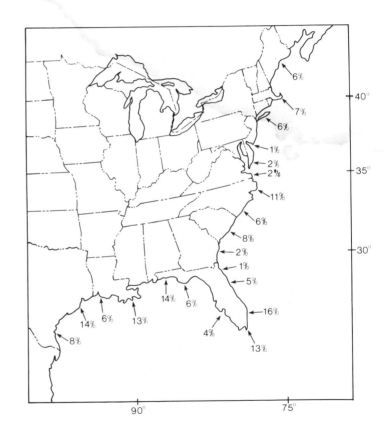

1.17 Probability (percentage) that a hurricane will occur in any one year for various locations on the U.S. East Coast and Gulf Coast. Although these values are low for Maryland and Virginia, probabilities rise sharply for the Outer Banks, just to the south in North Carolina. Modified from Simpson and Lawrence, 1971 (reference 13, appendix C).

The dynamic coast

Bay is surrounded by a yoke of land (the Delmarva Peninsula and the mainland south of the mouth of the Chesapeake Bay) over which any hurricane must pass before entering the Bay region. Hurricanes typically lose a great deal of their strength as they cross land, downgrading into tropical storms. While the tropical storm that is more common in the Chesapeake Bay region is less deadly than hurricanes, it too can be severely destructive. In 1972 Hurricane Agnes, a weak hurricane, was downgraded to tropical storm status as it entered the Chesapeake Bay region (fig. 1.18). Yet, persistent high winds generated storm surges and destructive waves. Torrential rainfall caused erosive runoff and induced cliff failure along the high bluffs of the western shore. This "downgraded" storm caused over $2.1 billion in damage, which makes it one of the costliest storms ever to hit the East Coast of the United States!

Historical records of storms show that in the recent past between one and two tropical storms affect the Bay each year, occasionally reaching hurricane strength (references 19, 20, and 25, appendix C). From 1667 to 1900 eleven "destructive" storms have been identified, which averages one storm every 20 to 25 years. This suggests that the average coastal dweller, and his home, will experience more than one destructive storm in a lifetime.

Impacts of storms on shorelines

The severity of the impact of a coastal storm on a segment of shoreline is not simply related to "bigger" or "smaller" storms.

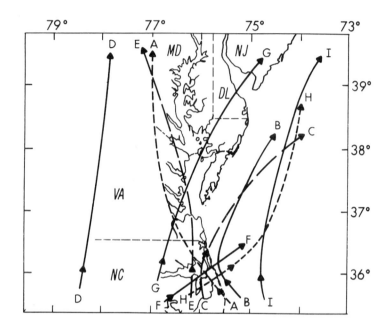

1.18 Twentieth-century hurricane tracks in the vicinity of the Chesapeake Bay and Virginia barrier island coast. Specific storms include (A) August 1933, (B) September 1936, (C) Barbara, August 1953, (D) Hazel, October 1954, (E) Connie, August 1955, (F) Flossy, September 1956, (G) Donna, September 1960, (H) Agnes, June 1972, and (I) Gloria, September 1985. Modified after Neumann et al., 1978 (reference 12, appendix C).

Several factors come into play, including the velocity at which the storm system migrates and its path. The forward velocity of the storm system can range from 0 (a "stalled" storm) to over 60 mph. The speed at which a storm moves has two contrasting impacts. The forward speed of the storm adds to the speed of the winds within the storm, increasing its destructive powers. On the other hand, the speed at which the storm moves limits the time available to generate waves or create wind surges. A storm system that charges through an area at 60 mph arrives quickly and passes, allowing little time for the waves to build, or coastal floods to encroach further landward through successive high tides.

The path a storm system takes plays a role in determining not only the intensity of the storm, but what shoreline will be most severely affected. If a hurricane makes landfall in the area of Chesapeake Bay, storm winds to the right (north or east) of the eye will be moving in the same direction as the storm center, resulting in more intense winds, larger waves, and higher surges to the right of the eye (fig. 1.19). A hurricane that tracks northward offshore of Chesapeake Bay generally causes southerly winds over the Bay, which pushes water to the south causing the largest storm surge to occur in the lower Bay (fig. 1.19). Hurricanes Connie (1955), Flossy (1956), and Donna (1960) were this type (fig. 1.18). Similarly, a storm that makes landfall south of the Bay along the North Carolina coast generally creates surges in the lower Bay and along the Western Shore. If the center of the hurricane moves landward of the Bay, water is pushed to the north causing maximum flooding in the northern (Maryland) portion of the Bay.

Storm paths, however, do not follow simple straight lines. Predictions of the paths that hurricanes will take are rarely accurate. The difference between a devastating hurricane and a moderate hurricane for a given location may result from a slight fluctuation in the path or speed of a storm. A good example is Hurricane Gloria, 1985, clearly one of the most intense hurricanes of this century, with wind velocities of over 150 mph. Coastal residents and communities prepared for potential destruction, but the storm moved seaward of most of the East Coast, minimizing impact (which was still significant).

Individual property owners are concerned not with the location of the highest surge, but rather with the effect on their immediate area. Fifty-nine flood-prone communities have been identified in the Bay, 32 of which are designated as "critical" with respect to a 100-year tidal flood (reference 14, appendix C). However, all low-lying coastal areas are susceptible to flooding under the right set of conditions. Neither a hurricane nor a nor'easter is requisite for an extreme flood. Because the Bay is extremely shallow, the wind is very effective at pushing and piling water along the shoreline. A steady wind blowing for a long period of time in the right direction, perhaps during a spring tide, may cause significant flooding in a given location.

A story of shoreline development

In 1866, after the Civil War had ceased to change the landscape of Virginia, a large stand of pines along the Eastern Shore felt

22 The dynamic coast

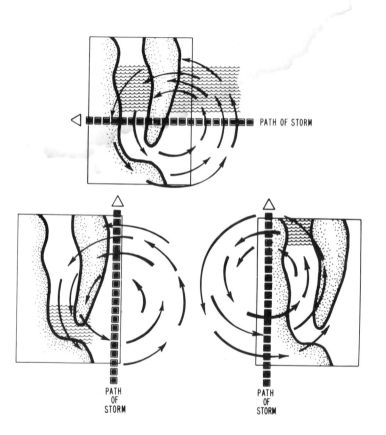

the effects of a quieter battle. This confrontation was waged not by the destructive weapons of man, but rather by the competing forces of Mother Nature. The pines stood on a peninsula of land appropriately named Savage Neck. They faced the winds and waves of Chesapeake Bay, which were ceaselessly changing the geography of Northampton County. Each tree became a temporary monument to this conflict, standing briefly at the water's edge before yielding to the Bay.

Around the turn of the century, by some accounts, Savage Neck's beautiful vista attracted a group of investors. Taking advantage of the remaining trees, beach, and Bay, they constructed a modest hotel in a location with the choicest view. The venture was quite successful. The investors were not oblivious to the changes in the shoreline, but they felt little concern—a substantial fastland remained between the hotel and the beach.

Later, after air travel had emerged as a major mode of transportation and national defense, the need arose for a strategically located air-navigation facility. Such a facility had very specific siting requirements. One of the primary concerns in site selection was a lack of physical obstructions, including trees. After proper deliberation, the Federal Aviation Administration chose an open

1.19 Relationship between the approach of a storm and flooding in Chesapeake Bay. Circular arrows show wind direction, while wave patterns indicate area of maximum flooding. Modified after Pore, 1960 (reference 19, appendix C).

field on Savage Neck facing Chesapeake Bay and constructed the facility called a VOR.

Before and after the VOR came into existence, private individuals were attracted to the sandy beaches and spectacular Bay view at Savage Neck. Over time summer cottages were built and those who occupied the cottages used the unobstructed beach for all forms of recreation including long walks. Many of these beachcombers were attracted to a section of the beach down by the VOR. Here they could find a number of interesting objects, aside from the normal flotsam and jetsam associated with a tidal shoreline. In the shallow water they found red bricks, pieces of pottery, old bottles, and occasionally rusted hunks of metal that might have been a tool, a door hinge, or some other artifact to fire the imagination.

Each spring the cottage owners faced the realities of change when they inspected the winter storm damage. Usually the change was recorded with the loss of a favorite bush or the steps to the beach. Occasionally, there was a reverse trend with a significant gain in the width of the beach. Sometimes the loss of beach was alarming. The only common denominator was that all experienced change.

With time the loss of fastland mounted and there was considerable discussion as to how to combat the problem. In the end the only feasible solution was to move the cottages landward a "safe" distance. Safe was considered to be 300 feet. Afterward the shoreline remained relatively stable for a time, discounting minor annual losses. But this apparent lull in shoreline retreat was dramatically broken with the "Ash Wednesday" storm of March 1962. Over three frightening days the unprotected bank yielded 40 to 60 feet to the Bay. Afterward the shoreline seemed to become more unstable and the annual rate of loss increased to over 10 feet per year. It became apparent that some measure of protection was necessary.

The decision was made to stop the erosion of the shore. A number of efforts were tried, including a concrete seawall, a groin field, and concrete rubble. Despite these efforts (and expenses) the shore continued to retreat, and one home that could not be moved was lost. Subsequently, an experimental structure was built with large sand-filled PVC-coated nylon bags. The bags were installed off the seaward end of the previously installed groins in a low, single-layer row running parallel to the shore before abutting the toe of the bank. The purpose of this structure (called a *sill*) was to "*perch*" the beach by serving as a low miniature breakwater, reducing the wave energy in the lee of the structure, and increasing the residence time of the sand. Under the right conditions, the incoming waves would push the sand to a higher elevation at the toe of the bank, thus preventing storm waves from reaching the bank. The sill experiment was a partial success. It perched two to four feet of sand above the normal backshore elevation. As a result, the cliff remained untouched by storm waves for nearly five years. A small vegetated dune formed at the base of the cliff.

Unfortunately, beginning with a modest nor'easter in 1977, storm waves overtopped the small dune and again began carrying bank material into the Bay. The sill had slowed the erosion rate, but it was only a matter of time before it became ineffective. Once the waves eroded an additional 20 feet of bank, the sill was completely unable to perch a beach. Adding to the problem was the deterioration of the bags themselves. Sand abrasion wore off the PVC-coating, exposing the nylon to degradation by sunlight. Debris carried by waves could then damage the bags.

Over a 13-year span various kinds of shore-protection schemes were tried. Although the sill was partially successful, it could not provide protection in the face of storm surges exceeding four feet. The net result was that the landowners once again faced the potential loss of their homes. After initial discouragement, they realized that their best choice was to move the houses once again. Today, the once-lucrative hotel on Savage Neck is a disappearing pile of debris in the nearshore and the VOR is under seige by the Bay. Its useful life may coincide with the arrival of the shoreline at its front door.

The moral of the story

This story of one small community's shoreline experience illustrates universal truths for all the shores of Maryland and Virginia, whether in Chesapeake Bay or along the Atlantic Coast. The beauty of the shore attracted development, but site selection was made without an understanding of shoreline processes. As a result, dwellings were built in an area where rapid shoreline retreat was taking place. The actual observed rates of shoreline retreat varied from less than one foot per year to as much as 36 feet per month for one five-month period! So much for averages. The early residents were wise enough to retreat in the face of natural hazards. They built houses that could be moved, and they moved them rather than confront nature.

The purpose of this book is to help landowners avoid the pitfalls in living with the shore. Chapter 2 outlines shoreline types and how they respond to natural processes; chapter 3 outlines some fundamental truths of the shoreline and how to avoid becoming entrapped by shoreline-engineering structures that often fail to solve the erosion problem; chapter 4 provides site-specific information concerning Chesapeake Bay and allows you to evaluate the risk associated with most developed or developable Bay shores; chapter 5 addresses similar questions for Virginia Beach and Virginia's open-ocean, barrier-island shore; chapter 6 outlines legal aspects that are important to shoreline property owners and residents of Maryland and Virginia; and chapter 7 presents useful information on construction techniques for building or improving dwellings within the coastal zone. Finally, the appendixes provide safety information on hurricanes and storms (appendix A), other sources for information on living near the shoreline (appendix B), and useful references (appendix C).

2 Coastal environments

If you plan to live near or invest in coastal areas, you should understand the types of environments found there and the natural processes affecting them. By knowing something about the coastal environment you can often predict whether an area has erosion or flooding problems. There are certain types of coastal features that exist only in areas where wave erosion is negligible. Conversely, there are shore types that signal major erosion problems. This chapter describes the major shore types in Maryland and Virginia, identifies the processes that formed these environments, and discusses the stability and safety of each shore type. This information will allow you to visit a coastal site and evaluate potential problems. This could save you a substantial amount of money and frustration!

This book discusses two very different coastal environments. Within Chesapeake Bay, an estuarine environment, the shoreline is highly variable, composed of a range of shore types from flat beaches to high bluffs. The waves are smaller and less destructive than those attacking ocean-facing coasts. Conversely, the open Atlantic Coast is predominantly composed of sandy barriers and is exposed to much larger and more dangerous waves. This does not mean the erosion and flooding problems are less significant within Chesapeake Bay; some of the highest erosion rates along the East Coast occur in the Bay. Because of the differences between the Bay and the ocean environments, the two areas will be discussed separately. Nevertheless, there are some shore types that are common to both. When reading through chapter 2, it would be useful to keep this in mind. For instance, the marshes in lower Chesapeake Bay are similar to the coastal marshes located behind the Virginia barrier islands, and they were formed by similar processes.

Shore types within Chesapeake Bay

Bluffs: erosional scarps

A significant portion of the Chesapeake Bay shore is composed of bluffs, the eroded edge of the land. These bluffs range in height from a few feet to over 100 feet and are composed largely of unconsolidated materials. Property owners should be aware that these bluffs are also subject to failure. For purposes of discussion, bluffs are arbitrarily divided here into *high bluffs* (greater than 20 feet in height) and *low bluffs* (less than 20 feet in height). The best example of high bluffs occurs along the western shore in the area known as Calvert Cliffs (fig. 2.1). Low bluffs are common throughout the Bay and its tributaries.

2.1 High bluffs located at Calvert Cliffs, Maryland.

Failure or *slumping* occurs when the material composing the bluff collapses due to gravity. The result of slumping is a more gradual slope, which increases the bluff's stability. Unfortunately, wave action continues to remove material from the base of the bluff, which steepens the slope again, decreasing stability. Consequently, coastal bluffs rarely attain stable slopes. It may appear that the key to stabilizing the bluff is protecting the base from wave attack. This is clearly one important step to take, but not the entire answer. Many bluffs in the Bay with stabilized bases continue to erode.

Slumping of bluffs does not have to be caused by wave action alone. Freshwater runoff from the land can cause bluff failure independently of wave action. Most of the bluffs in the Bay are composed in part of finer silt and clay, which inhibits the flow of groundwater. Groundwater collects above these impermeable layers until it finds its way to the face of the bluff (fig. 2.2). Seeping groundwater on a steep bluff face tends to cause *spalling* (clumps of mud breaking off the bluff face). In addition, runoff from rainfall flows down the bluff face, eroding small channels. These effects can be reduced by *grading* (reducing the slope of the bluff) and by anchoring the soil by growing grass or other vegetation.

A giant slump in Westmoreland County, Virginia, occurred because of a clay layer at the base of a cliff. The cliff failed after a several-week period of heavy rainfall. Most likely the failure was because added water accumulated along the top of the impermeable clay layer, which weakened the bluff. The remnants of the slump have been terraced in an attempt to prevent future problems.

Unfortunately, many landowners will not deal with the erosion problem until the situation is critical. Attempts to stabilize a bluff are made only when the house is on the edge. At this point the owner cannot simply grade and vegetate the slope. The house must also be moved.

Marsh shorelines: a natural defense

Marshes are a common, and often the dominant, coastal feature in many areas of Chesapeake Bay (fig. 2.3). Marsh grasses and shrubs are similar to dune grasses: they thrive in a very hostile environment and they tend to collect sediment. During flooding, the plants baffle water currents and trap suspended material from the water. The resulting deposit is a muddy material held together in a dense framework of plant roots and rhizomes.

The types of marsh vegetation vary according to salinity and the frequency of flooding in the area. High-salinity marshes are dominated by relatively few types of plants. Among the most common are *Spartina alterniflora* (cord grass), *Spartina patens*, *Salicornia*, and *Borrichia*. In the lower-salinity and freshwater marshes, the vegetation is extremely diverse. Commonly found are *Scirpus*, *Hibiscus*, *Juncus* (needle rush), *Panicum*, *Phragmites*, *Typha*,

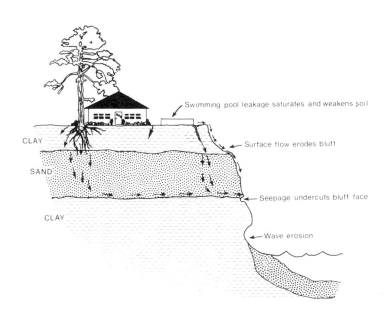

2.2 Infiltration of groundwater (small arrows) from various sources can lead to bluff failure. The presence of an impervious layer such as clay causes a buildup of groundwater which weakens the bluff. Modified after U.S. Army Corps of Engineers, 1981 (reference 76, appendix C).

2.3 Large marsh near Crisfield, Maryland.

Zinania (wild rice), *Carex* (sedges), *Distichlis* (salt grass), *Sagittaria* (arrowhead), *Baccharus, Iva, Lecrsia* (cut-grass), *Acorus, Fimbristylis, Peltandra* (arrowarum), *Pontederia* (pickerelweed), *Eleocharis* (spike rush), *Pluchea,* and *Rhynchospora.*

Marshes are an environment uniquely adapted to a rising sea level. As sea level rises, the marsh level increases by trapping mud and depositing plant debris, forming peat. Thus, peat deposits can become thick in subsiding areas such as the Bay. However, if the sea-level rise is faster than the marsh can accumulate sediment, then submergence of the marsh occurs. The marsh at Blackwater Wildlife Refuge on the Eastern Shore has lost over 50 percent of its area since earlier in this century due to submergence (fig. 2.4). Not all marsh systems are stable!

Marsh systems tend to develop in almost any intertidal area where wave energy is low. Two types of marshes occur within the Bay. Extensive marshes covering many acres form in flat areas and at the heads of creeks, where waves are negligible. There are over 200,000 acres of marsh in the Virginia portion of the Bay alone! The shoreline of these areas are referred to in this book as *marsh margins* or simply marshes. In some cases these shores are fronted by beaches or sand barriers (fig. 2.5). A second type of marsh develops in thin strands in front of the fastland (fig. 2.6). This shoreline type is found in protected areas. It is known as a *fringe marsh* and is an indicator of shoreline stability. If a shore is eroding rapidly, there is no fringe marsh. Be careful though. Even a fringe marsh may be retreating if a scarp has formed. The presence of a scarp indicates the shore may no longer be stable and is actually eroding.

Marshes play several roles in coastal systems. Marsh peat is a dense, cohesive material that resists erosion by wave action. Marsh plants also dissipate wave energy, protecting the shoreline behind the marsh. Waves do gradually erode the marsh edges, however, and can erode a marsh very rapidly. The amount of peat lost by wave attack in a single storm may have taken years to form. In fact, some of the highest retreat rates along the Eastern

2.4 Blackwater Wildlife Refuge, Maryland. Low sediment accretion rates have caused nearly half of these beautiful marshes to be lost since earlier in this century due to submergence.

Shore occur in the marshes. Still, as long as the marsh exists, the landward area will be protected against wave attack.

Extensive marshes are also a natural form of flood protection. Floodwaters moving landward through marsh are slowed down, reducing the flood height on the adjacent shore. While most landowners would prefer a sandy beach over a fringe marsh along their shoreline, a marsh is often the cheapest form of shore protection available.

In some areas of the Bay, sea grasses below the low-tide line have an effect similar to marsh plants. Although most effective in sheltered or low-wave energy areas, planting of sea grasses may be a useful tool for reducing shoreline erosion.

Estuarine beaches: the wave buffer

Estuarine beaches described in this book are accumulations of sand fronting the fastland. These beaches are the result of sediment moving from eroding shorelines. Silt and clay are readily carried off by waves, leaving sand behind to form a beach (fig. 2.7). Most sand comes from the area behind a beach that has been eroded by waves. The sand may be from the immediate vicinity of the

2.5 Marsh fronted by beach south of Gwynn Island, Virginia, above.
2.6 Small fringing marsh in Accomack County, Virginia, below. Marsh vegetation helps protect the shoreline from erosion and often indicates shoreline stability.

2.7 Estuarine beach located along the Choptank River, Maryland.

beach, or it may have been carried along the shore by longshore or littoral currents. Ironically, the natural buffer against erosion —the beach—depends on erosion for its construction material—sand. Unfortunately, the sediments being eroded along much of the Chesapeake Bay shore are not sandy. The results are few or no estuarine beaches. Other sources, such as the onshore movement of sand, are limited. Also, much of the sand carried by the rivers draining into Chesapeake Bay is trapped within the rivers.

Many factors affect the size of an estuarine beach, but the amount of sandy material fed in from nearby sources is critical. Beaches adjacent to features such as sandy bluffs that are eroding tend to be wider and longer. Conversely, beaches, if present at all, are small when they are adjacent to eroding marshes or clay bluffs, which provide little sand. The material in an eroding bluff also plays a role in how the beach buffers the waves. A beach adjacent to an eroding sandy bluff will have a thicker vertical accumulation of sand. A beach adjacent to an eroding clay bluff will have only a thin veneer of sand overlying the clay.

Estuarine beaches composed of a thin veneer of sand lying on clay substrates are very susceptible to erosion. When waves break on the shore, the water surges up the beach (*swash*), then gravity carries the water back downslope (*backwash*). As the swash runs up the beach, some of the water percolates down into a sandy beach, so less water is left for the backwash, and less erosion takes place. If there is a clay layer beneath the beach, less water can percolate downward, so more water is in the backwash carrying more sand seaward. Sand carried seaward by the backwash may be lost to the beach.

Most beaches have gentle slopes and are an important natural shore defense against wave erosion. Under normal conditions, beaches dissipate much of the wave energy before it can cause erosion to the mainland. Severe erosion takes place in the Bay when waves break at the back of the beach at times of elevated water levels, e.g., storm surges. Waves then break on bluffs, dunes, or houses. In general, the higher and wider the beach, the greater the protection from wave attack.

Thick, sandy estuarine beaches are better buffers to wave attack and offer greater protection to the shore than do impermeable, clay-based beaches. Unfortunately, such impermeable beaches are very common in the Bay. In addition, less sand builds up on these beaches, which causes them to be lower in elevation and afford less storm protection.

Barrier beaches and spits

Barrier beaches are long, sandy beaches and dunes separated from the mainland by lagoons or marshes. They are typically found in the lower Bay. Where these barriers have remained unstabilized, they tend to "roll-over" the salt marshes behind them. Very frequently, the salt-marsh sediment is exposed on the seaward side. This exposed marsh may be recolonized by grasses to form a fringe marsh.

Although the narrow barrier beaches are eroding at rates comparable to the rest of the shore, they are not disappearing. The packet of sand that forms the barrier beach migrates landward through wind transport and storm overwash. Such migration is similar to the movement of barrier islands discussed in the next section.

Spits are elongated beaches formed from the longshore transport of sand. Spits typically form when there is a sharp bend in the shoreline or across the entrance to small embayments. The low supply of sand in the littoral system in the Bay causes spits to be relatively small.

Dunes: stored sand

Where sand supplies are abundant, the wind may transport sand inland from the beach and store the sand in the form of dunes (fig. 2.8). The dunes are an integral component of the coastal environment because they act as a reservoir of sand that may be transported back to the beach. In addition, the dunes act as a buffer against waves and floodwaters.

Few plants can thrive in the dynamic environment of the dunes, which are exposed to salt spray, periodic flooding, and shifting sands. The dune grass *Ammophila,* Greek for "friend of sand," however, fills this niche. For *Ammophila* to thrive and grow, it must be buried by sand. This burial starts the cycle of coastal dune growth. As the plant grows it traps wind-blown sand, which further stimulates growth. The plant can grow through a burial of up to three feet per year. Its dense root system (*rhizomes*) holds sand in place, stabilizing the dunes.

Along the shoreline of the Grandview Nature Preserve in Hampton, Virginia, *Ammophila* has successfully built up dunes to an elevation of four feet. The presence of dunes here has reduced the frequency of beach sand loss due to storm overwash, in turn reducing the rate of shoreline erosion.

Ironically, though *Ammophila* has adapted to exist in the harshest of environments, it has no tolerance to being walked on or driven over. This quickly kills the plants, and the loose sand is dispersed by the wind. The loss of the dunes increases the shore's susceptibility to flooding or erosion.

Because of the high silt and clay content of the fastland, sand is a scarce commodity in the Bay, so dunes tend to be less common and smaller than in other coastal areas. Where dunes do occur, a small amount of protection for the dune plants gives a large amount of shore protection. Where dunes are not present, dune grass can be planted in sandy areas to create dunes. The plants will be self-sufficient in a few seasons.

Lowlands

Maps and air photos show some land areas that slope gently to the water's edge (fig. 2.9). Such areas, which are narrow with little beach or fringing marsh, do not fall conveniently into the shore categories of bluff, beach, marsh, dune, or spit and are grouped here into the general category of *lowlands*. These lowland areas often suffer high rates of erosion and are frequently flooded.

Man-modified shores: artificial stabilization

Man's response to shoreline erosion and coastal flooding often is to build shore-stabilization structures (see chapter 3). In some

2.8 Dune fields located at Deal Island, Maryland, above, and Russell Island, Virginia, below. The dunes at Deal Island appear to have undergone relatively recent erosion as evidenced by scarps. Conversely, more stability is indicated at Russell Island by the presence of fringing marsh grass.

Coastal environments

2.9 Extensive lowlands located near the mouth of the Choptank River. Many of these lowland deposits are rapidly eroding, especially where exposed to large fetches. Note the turbid water which surrounds the land. Resuspension of nearshore bottom sediments and shoreline erosion create these turbid plumes even during moderate wind conditions.

regions, particularly commercially developed or urban areas, such structures obscure the natural shore type. The presence of structures designed to prevent or reduce erosion generally can be taken to indicate that there is a problem (e.g., erosion or flooding). In such areas, development should be approached with caution.

The open-ocean coast

The entrance to Chesapeake Bay and the coastal areas facing the open Atlantic Ocean in Virginia are composed largely of barrier islands backed by salt marshes and lagoons (fig. 2.10) or sandy beaches which have formed directly in front of the mainland. These environments are extremely dynamic and are continuously changing. Most of the barrier islands located north of the mouth of Chesapeake Bay are owned by federal and state agencies or private conservation foundations. Consequently, little development will occur on these beautiful barriers. In some isolated instances, development for residential use has been proposed. In contrast, much of the open-ocean coast south of the entrance to Chesapeake Bay (e.g., Virginia Beach) is very heavily developed. Here a major confrontation between man and nature is taking place. Whether you are interested in living on an open-ocean coast or are a curious visitor, understanding the origin and processes shaping the shore environment will enhance your enjoyment.

The origin of barrier islands

The forces acting on the islands today are the same ones that originally created them. Although the origin of barrier islands has been long debated, it appears that the Delmarva barriers formed because of events that occurred out on what is now the continental shelf. Approximately 15,000 to 20,000 years ago, when sea level was as much as 150 to 300 feet lower than today, the rivers flowed

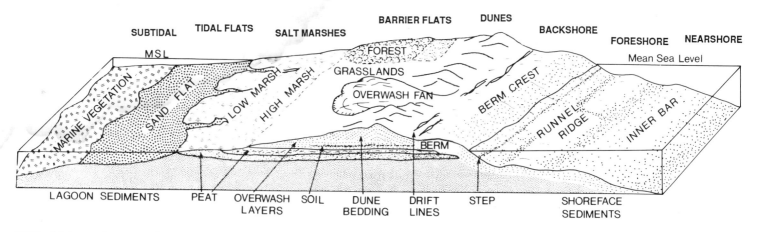

2.10 Major environments found along Delmarva barrier islands, above, and Smith Island, Virginia, right. Arrow on seaward side of barriers shows waves breaking on offshore bar. Arrow on landward side of lagoon shows washovers moving into salt marsh.

across the continental shelf eroding valleys. When sea level rose, the largest of these valleys formed Chesapeake and Delaware bays. Other smaller valleys formed smaller embayments. If this had been all that occurred, the shoreline today would be jagged. Nature, however, tends to straighten jagged shorelines. Waves attacked the headlands (the areas of land that protruded seaward between flooded valleys) and built spits extending from the headlands across the embayments (fig. 2.11). As sea level continued to rise, the low-lying land behind such spits, plus the sand dunes of the old headland shorelines, then became flooded. The flooding behind the old dune-beach complexes resulted in their detachment from the mainland, and the barrier islands were born.

Barrier islands: islands on the move

You might ask why the newly formed islands were not covered by the sea if sea level continued to rise. The answer is this: When the level of the sea is rising, barrier islands refuse to stay in one place. They migrate toward the mainland. The more rapidly sea level rises, the more quickly the islands move. For the islands to remain islands, needless to say, the mainland shore moves too. Otherwise

Accelerating sea-level rise

As discussed in chapter 1, recent studies suggest an acceleration in the rise in sea level. Sea level is now rising at a rate of perhaps slightly more than one foot per century. Keep in mind that this refers to a vertical rise. The horizontal change—the distance islands migrate as a consequence—is much greater. How much a specific island moves depends on the slope of its migration surface: the gentler the slope, the more the island will migrate.

Barrier-island migration

Do you want to prove to yourself that barrier islands migrate? If you're standing on one now, walk to the ocean-side beach and look at the seashells. Chances are that on most Delmarva beaches you will find shells of oysters, clams, or snails that once lived in the area on the back side of the barrier island toward the mainland. How did shells from the back side get to the front side? The answer is that the island migrated over the back-side area, and waves attacking and breaking up the old back-side sands and muds threw the shells up onto the present-day beach. (This assumes you are looking at a natural beach where sand has not been pumped in from behind the island as artificial nourishment.) In addition, storms occasionally expose salt-marsh peats (which formed on the back of the islands at some earlier time) on present-day ocean-side beaches. On Whale Beach, New Jersey, a patch of mud that appeared on the beach after a storm contained (much to the surprise of some beach strollers) cow hooves and fragments of colonial

the islands would run aground. This is indeed what happened: the shoreline on the mainland retreated as the sea advanced over the land (fig. 2.12).

The sea-level rise was quite rapid until about 5,000 years ago, at which time it slowed down considerably. Hence, up until 5,000 years ago Delmarva's barrier islands were moving landward at an impressive clip. Such rapidly moving islands tend to be low, very narrow strips of sand.

When the slowdown in sea-level rise came, many islands stopped migrating altogether. Their relatively stable position allowed sand from various sources to accumulate, and the islands began to widen. This relative stability, however, has recently come to an end.

36 Coastal environments

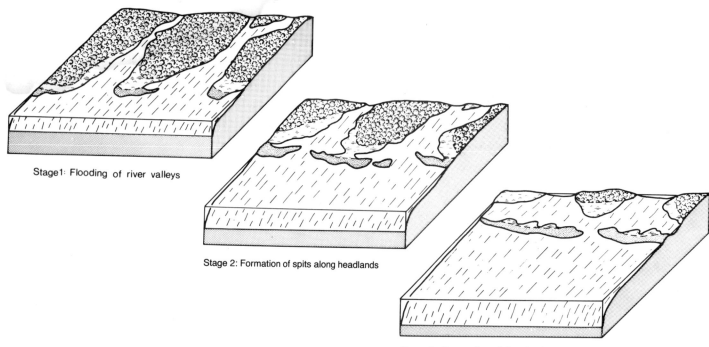

Stage 1: Flooding of river valleys

Stage 2: Formation of spits along headlands

Stage 3: Separation of barrier from mainland

2.11 Stages of barrier development during the Holocene sea-level rise.

Coastal environments   37

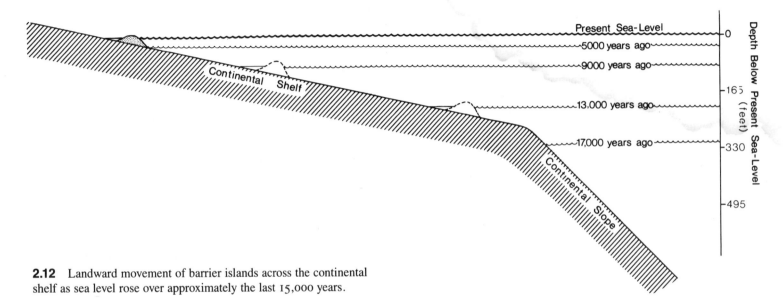

2.12 Landward movement of barrier islands across the continental shelf as sea level rose over approximately the last 15,000 years.

pottery. The mud was formerly salt marsh on the back side of the island where a colonist had dumped a wagonload of garbage. Since colonial times this particular section of the island had migrated its entire width!

If you haven't guessed already, *island migration* is the term that geologists use for what beach-cottage owners call *beach erosion*. In order for an island to migrate, the side on the ocean (the *front side*) must move landward by erosion, and the side toward the mainland (the *back side*) must do likewise by growth. As it moves, the island must somehow maintain its elevation and bulk.

The beach on the front side moves back because the sea level is rising. The rise in sea level is the main cause of beach erosion worldwide, although man can make erosion worse, sometimes much worse, by interfering with the sand supply.

There are several ways for an island to migrate or widen in a landward direction. On the narrow, low Delmarva barrier islands that are separated from the mainland by an open body of water (called a *sound* or *lagoon*) as at Chincoteague, Magathy, or Metomkin Bay (fig. 2.13), a widening process is that of incorporation of old flood-tidal deltas from closed inlets.

An *inlet* is the channel of water between adjacent barrier islands (fig. 2.14). The inlet may be a permanent feature or a short-lived feature that forms when water breaches the island during a hurricane. Eyewitness accounts and other evidence confirm that water sometimes breaks through from the sound side, or back side, of the island as the storm subsides and high tides retreat. Over the years following the formation of an inlet, sand carried by incoming tides pours through the gap and into the sound. This mass of sand is called a *flood-tidal delta*. There is also a delta called an *ebb-tidal delta* on the ocean side of an inlet. This is formed by the interaction of currents flowing out of the sound during falling tide and waves.

If an inlet closes or migrates away from the flood-tidal delta, salt marshes sometimes establish themselves. Salt-marsh grasses trap sediment and cause the land to be built up almost to high-tide level so that new land is added to the back of the island. The inlet's former position is marked only by a marsh bulging into the sound.

When an inlet migrates laterally to a new position, the flood-tidal delta moves with it. In other words, as the inlet migrates, sand continues to pour into the sound, and a series of new flood-tidal deltas are formed along the entire area of inlet migration. In this way the island widens over the full distance that the inlet shifts.

2.13 Lagoonal system (Metomkin Bay) located landward of Metomkin Island, Virginia. Extensive washover features (arrows) occur on the landward side of the low barrier. Photograph by Kenneth Finkelstein.

Coastal environments 39

2.14 Typical southern Delmarva tidal inlet (Wachapreague Inlet and associated shoals).

Another way that islands—especially narrow ones—migrate landward is by direct frontal *overwash* of storm waves from the ocean side (fig. 2.13). Most barrier islands receive overwash during storms. On large barriers the overwash may barely penetrate the first line of dunes. On low, narrow barriers, overwash may be carried across the island to reach the sound. Overwash waves carry sand that is deposited in tongue-shaped or fan-shaped masses called *overwash fans*. When such fans reach into the sound or into the marsh, the island is widened. If islands are backed by salt marsh, overwash sediment may bury the marsh, thus nourishing it. Interestingly enough, many islands need overwash to survive.

Ocean-facing beaches: the dynamic equilibrium

Open-ocean beaches are one of the earth's most dynamic environments. The beach—or zone of active sand movement—is always changing and always migrating, and we now know that it does so in accordance with the earth's natural laws. The natural laws of the beach control a beautiful, logical environment that builds up when the weather is good, and strategically (but only temporarily) retreats when confronted by big storm waves. This system depends on four factors: waves, sea-level rise, beach sand, and the shape of the beach (fig. 2.15). The relationship among these factors is a natural balance referred to as a *dynamic equilibrium:* when one factor changes, the others adjust accordingly to maintain a balance. When we enter the system incorrectly—as we often do—the dynamic equilibrium continues to function but in a way that may be harmful.

It is important to keep in mind that the *active beach* extends from the toe of the dune to a water depth of 30 to 40 feet offshore. The part on which we walk is only the *upper beach*.

How does the beach respond to a storm?

Old-timers and storm survivors have frequently commented on how flat and broad the beach is after a storm. The flat beach can be explained in terms of the dynamic equilibrium: as wave energy increases, sand is eroded from the beach, changing its shape. The reason for this response to storms is logical. The beach flattens

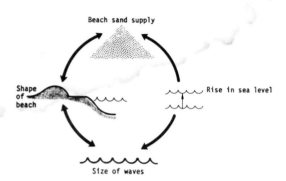

2.15 Factors affecting the dynamic equilibrium of a coastal system.

itself so that storm waves will expend their energy over a broader and more level surface. On a steeper surface storm-wave energy would be expended on a smaller area, causing greater damage. Figure 2.16 illustrates the way in which the beach flattens during a storm. In summary, the waves take sand from the beach or the first dune and transport it offshore. If a hot-dog stand or beach cottage happens to be located on the first dune, it may disappear along with the dune sands.

Sometimes, besides simply flattening, a storm beach also will develop one or more offshore bars. The bars serve to trip the large waves long before they reach the beach. The sandbar produced by storms is easily visible during calm weather as a line of surf a few to tens of yards off the beach.

An island can lose a great deal of sand during a storm. Much of it will come back, however, gradually pushed shoreward by fair-weather waves. As the sand returns to the beach, the wind takes over and slowly rebuilds the dunes, storing sand to respond to nature's next storm call. In order for the sand to come back, of course, there should be no man-made obstructions—such as seawalls—between the first dune and the beach. Return of the beach may take months or even years.

How does the beach rebuild after a storm?

Beaches grow upward and seaward in several ways, principally by (1) bringing in new sand by the so-called longshore (surf zone) currents, or (2) bringing in sand from offshore by forming a ridge and runnel system. These two ways of beach widening are not mutually exclusive.

Longshore currents are familiar to anyone who has swum in the ocean; they are the reason you sometimes end up somewhere down the beach, away from your towel. Such currents result from waves approaching the beach at an angle, which causes a portion of the energy of the breaking wave to be directed along the beach. When combined with breaking waves, the current is capable of carrying large amounts of very coarse material for miles along the beach. A current flowing offshore created by waves is referred to as *undertow*.

Ridges (sandbars) and *runnels* (low areas in front of sandbars) form following storms and virtually march onto the shore and bring

Coastal environments 41

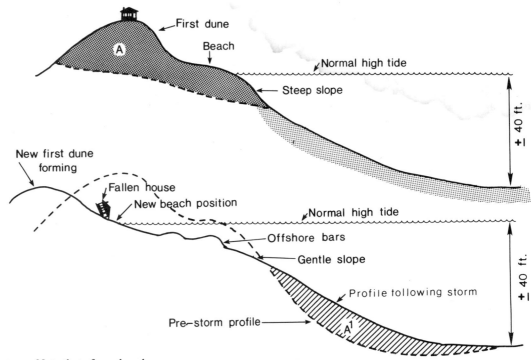

2.16 Changes in a beach following a storm. Note that after a beach has been eroded by a storm (lower schematic) the dunes and beach have receded and the slope of the beach is gentler. Some of the sediment eroded from the dunes is deposited in the offshore bars or farther seaward.

(Shaded area A^1 is approximately equal to shaded area A.)

sand to the beach. The next time you are at the beach, observe the offshore ridge for a period of a few days and verify this for yourself. You may find that each day you have to swim out a slightly shorter distance in order to stand on the sandbar.

At low tide the beach frequently has a trough filled or partly filled with water. This trough is formed by the ridge that is in the final stages of welding onto the beach. Several ridges combine to make the *berm* or beach terrace, on which sunbathers loll.

The long-range future of beach development

The long-range future of beach development depends on how individual communities respond to their migrating shorelines. Those communities that choose to protect their front-side houses at all costs need only look to portions of the New Jersey shore to see the result. The life span of houses can unquestionably be extended by "stabilizing" a beach (slowing the erosion). The ultimate cost of slowing erosion, however, may be the loss of the beach. How long the destruction will take is highly variable and depends on shoreline or island dynamics.

If, when the time comes, a community grits its teeth and moves the front row of buildings or lets it fall in, the beach may be saved. So far in America, the primary factor involved in shoreline decisions, which every beach community must sooner or later make, has been money. Poor communities let the island roll on. Rich ones attempt to stop it.

Erosion on the back sides of our islands

Most of the Delmarva's barrier islands north of the Bay entrance have healthy salt marshes growing on the sound side, protecting the back sides from erosion. If bluffs or stumps appear at the water's edge on the back side, however, beware: erosion is occurring.

Rates of erosion for most back-island beaches have not been determined, but are often not trivial! The U.S. Army Corps of Engineers has determined that the erosion rates in the large lagoonal system behind Sandbridge and the barriers to the south are up to 10 feet per year—a rate equivalent to some on the ocean side. Back Bay has the highest erosion rate in Currituck Sound north of Oregon Inlet. It appears waterfront property *anywhere* can have its problems!

3 Shoreline engineering: stabilizing the unstable

For all practical purposes we can say that most of the shoreline of Chesapeake Bay and the open-ocean coastline of Virginia is eroding. A fundamental cause of this widespread erosion is the rise in sea level currently being experienced by all of the world's oceans (discussed in chapter 1). With or without a sea-level rise, however, it is likely that the Bay's shores would be retreating.

Some of this erosion is truly spectacular. Parts of Tangier Island have moved back well over one-half mile since the first accurate shoreline map was made in 1850. At a site near the mouth of the Potomac (west of Cod Creek), the long-term rate of erosion is 10 feet per year. In Lancaster County, Virginia, north of Windmill Point at the mouth of the Rappahannock, the long-term rate of erosion is eight feet per year. In 1983 a shoreline erosion problem in Hooperville on Hooper Island, Maryland, made local headlines when the edge of the oldest cemetery on the island was only 17 feet from the Bay. According to the memory of one long-time island resident, there was once almost a half mile of land between the shoreline and the cemetery, most of which was a corn field.

This shoreline retreat has caused the inevitable problem of destroyed buildings, docks, seawalls and lost farmland. Naturally the common response is to try to stop the erosion, or in engineering jargon, to *stabilize* the beach so it no longer erodes. Any procedure carried out or structure built for the purpose of halting shoreline retreat falls into the category of *shoreline stabilization* or *shoreline engineering*. Methods of stabilization range widely in cost, performance, and success. At one end of the spectrum is the introduction of salt-marsh grass in front of an eroding bluff. The other extreme in terms of permanence, expense, and flexibility is construction of a massive concrete seawall in front of the same bluff.

Open-ocean shorelines such as on Virginia Beach and Sandbridge are very dynamic and their retreat is very difficult to prevent. Many millions of dollars have been spent along a relatively short stretch of Virginia Beach to pump or truck in fresh sand and build various kinds of seawalls. Maintenance of a broad beach is imperative to the survival of the Virginia Beach tourist economy and large expenditures can be justified, at least in the minds of business leaders, to preserve both beaches and buildings next to the shore.

The task of battling shoreline erosion in Chesapeake Bay is much less formidable than along open-ocean shorelines simply because the average daily wave height is much smaller. In addition, less value is attached to the recreational beach in Chesapeake Bay than along resort areas such as Virginia Beach or Ocean City. Although many thousands of people swim daily in the Bay (except when the jellyfish abound), most swimming goes on in small groups at small beaches. Much less money is spent on erosion control if preservation of the recreational beach is not considered to be a high community priority.

The variety of devices used to halt or impede shoreline retreat is truly mind boggling. Rubber tires, offshore breakwaters of ingenious construction, sunken barges, and plastic imitation seaweed have all been used in attempts to break up waves and cause beaches either to build out or at least to stop retreating.

If you have decided that shoreline erosion threatens your building, you must follow certain rules to save yourself time, money, and headaches.

Rule 1: Know your options. Faced with an erosion problem, a homeowner may have a virtual blizzard of alternatives to consider. The possible options that a homeowner can consider fall into six broad categories. The options are introduced here and discussed in more detail later in this chapter. These options are:

1. Do nothing. This means let the shoreline roll on and if your house goes, so be it. The advantage to this option is that the beach and the associated natural-shoreline environment remain unharmed.
2. Construction setback. This is an option for all new structures. If you know there is an erosion problem, build back a safe distance from the shore.
3. Move threatened buildings. This has been done in Chesapeake Bay here and there, especially from atop high bluffs. The advantages are that the natural shoreline environment is left unharmed and the cost of construction and maintenance of walls, etc., is avoided.
4. Natural solutions. Marine and terrestrial grasses stabilize many undeveloped shores. In some instances transplanting the appropriate grasses can be used to reduce erosion.
5. Soft solutions. If the method of halting shoreline erosion does not employ construction of in-place objects, it is called a soft solution. In Chesapeake Bay the two most common soft solutions are beach nourishment and vegetative control. These have the advantage of maintaining close to a natural-shoreline system.
6. Hard solutions. Construction of something immovable constitutes a hard solution. These fall into two categories: (1) structures built perpendicular to the beach (groins and jetties); and (2) structures built parallel to the shoreline (seawalls, bulkheads, revetments, and breakwaters).

Rule 2: Talk with your neighbor. One of the major problems in shoreline stabilization is that everybody has a better idea about how to stop erosion. As a result, on many if not most stretches of shoreline where houses hug the waterfront, we have a hodgepodge of various types of structures holding back the sea plus some homesites where the owners have chosen to do nothing (fig. 3.1).

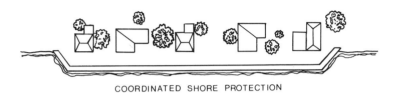

COORDINATED SHORE PROTECTION

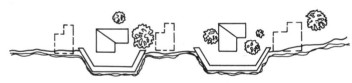

UNCOORDINATED SHORE PROTECTION

3.1 The effects of uncoordinated shore protection structures along a coastal reach. When neighbors cooperate, more successful erosion control is often achieved. In areas where different approaches or no structures at all are used, harmful interactions can occur. Modified from U.S. Army Corps of Engineers, 1981 (reference 76, appendix C).

Lack of a uniform or at least a coordinated approach is a real problem for shoreline property owners for a number of reasons:

1. In areas where there is significant longshore transport of sand, shoreline-stabilization structures (e.g., groins, jetties, or breakwaters) can halt the supply of sand to adjacent beaches and hence increase the rate of erosion there. Thus, erosion rates of adjacent properties where the owner chooses to do nothing may be greatly increased. Also "*end-around*" effects (i.e., greatly increased erosion rates at the end of walls) are well documented (fig. 3.2).
2. Property owners often put in structures at different times without planning for or understanding how they will interact with structures already built.
3. Even if all the neighbors put in their erosion-halting structures at the same time, they often will have different ideas or will have received differing advice concerning how to go. The result is a smorgasbord of structures, often with damaging interactions.
4. Your neighbor may not spend much money for shoreline protection. The failure of an inadequate wall may cause problems and more expenses for everyone in the neighborhood.
5. Once you've installed the shoreline-protection structure, you haven't solved the problem. Maintenance is a must, and if your next-door neighbor doesn't bother to walk down and check the seawall, or whatever, from time to time, you both may be in trouble.

For all of these reasons and even more, you should talk with your neighbor and coordinate all efforts to stabilize the shoreline. You need to talk with everyone along your reach of shoreline. A *reach* is a segment of shoreline, ranging in length from a few yards

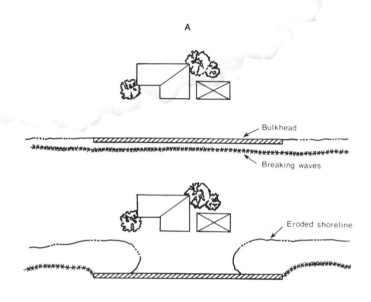

3.2 (A) Coastal structures such as bulkheads or revetments which are not properly tied in to the fastland often suffer end-around effects (erosion at the ends of the walls) which can lead to failure of the structure. The structure should be tied into the fastland as shown in the lower portion of the figure. Modified from U.S. Army Corps of Engineers, 1981 (reference 76, appendix C). (B) Scour at the end of a revetment located in lower Chesapeake Bay is causing weakening of the structure.

to miles, where shoreline processes and materials are interrelated. In the area of a reach, steps taken by one neighbor to halt erosion may have an impact on other neighbors. For example, thousands of bluffs and cliffs are eroding in Chesapeake Bay and the sediment derived from erosion ends up on nearby beaches. If your neighbor decides to bulkhead the base of the bluff to keep a house from falling in, the sand supply for your beach may be cut off and your beach will disappear. Similarly, if your updrift neighbor decides to build a groin, his groin may cut off the flow of sand to your beach. While his beach builds out, your beach erodes. Consequently, you must build a groin, passing the problem on to the next neighbor down the line. And on it goes. It is not always easy to tell what reaches of the Bay shoreline are interrelated and expert help may be needed.

Rule 3: Know your structure. An economic and an environmental price is paid whenever a shoreline-erosion structure is built. You must clearly understand construction costs, maintenance costs, and replacement costs. At the same time you must understand the likely environmental cost. Building a seawall or bulkhead may destroy the beach fronting the structure. Is it worth it? The most common types of structures and their impacts are discussed in this chapter.

Rule 4: Consider the long range. When deciding what to do about an eroding shoreline, you should consider your proposed solution in a long-term sense. Solving the erosion problem by a means that will only last a year or two may be self-defeating and ultimately very costly. Furthermore, when estimating dollar costs, consider both maintenance and replacement costs. Every shoreline-erosion device is designed to fail. That is, a certain "design life" is assumed after which it must be replaced. Of course, if a big-enough storm comes the day you finish construction, your efforts may disappear overnight. Look at long-range environmental impacts and long-range costs and *then* decide if it's worth it.

Solutions to coastal erosion

Humans have occupied and used the coast since time began and will continue to do so. The key to coastal use is the prevention and mitigation of potential problems. Coastal engineering as a science has advanced far enough to predict and prevent most problems dealing with coastal development projects. However, the average citizen, who knows little or nothing about coastal processes and engineering, should adopt a hazard-prevention policy. This policy is best accomplished by a "safe" construction setback, that is, build far enough back from the water's edge so that the structure will not be in danger from erosion or flooding for the lifetime of the structure. This policy accommodates coastal changes with little or no threat to nearby structures. Where a safe setback is not possible or construction already exists, other options of erosion protection are available. As discussed, they range from doing nothing or nonstructural, natural solutions to extensive structural measures or hard solutions. Several of these "solutions" to coastal erosion are presented in table 3.1 and discussed in the remainder of this chapter.

Do nothing

The do-nothing alternative involves total prohibition by local government of the use of any procedures or structures to prevent or slow down shoreline erosion. Thus, individuals who own buildings adjacent to a moving shoreline must eventually move back or demolish their beachfront structures. The advantages to this approach are that it involves no cost to the taxpayers and it preserves the recreational and aesthetic values of the beach. It also greatly reduces future imprudent development. Of course, this all results in a high cost to the individual property owner.

Construction setback

The most obvious way to avoid a hazard is to stay away from it! This logic equally applies to an eroding shoreline or unstable bluff. The prudent planner, recognizing the likelihood of some erosion or sliding over the years will build well back from the shore. How far back is a "safe" building setback? That question is difficult to answer and will vary from place to place. When in doubt, ask questions of neighbors, longtime inhabitants, town officials, or if necessary a coastal geologist.

There are a number of advantages to observing construction setback lines. The threat of erosion or flooding is greatly reduced when construction is back from the shore. The absence of construction allows natural shoreline processes to operate without interference and preserves the recreational and aesthetic values of the beach. In addition there are no long-term maintenance costs or permit problems.

The disadvantages of this approach are that water views are impeded, the lot must be deep enough for suitable setback, and erosion or landsliding is not stopped. Furthermore, the shoreline will probably be lapping at the houses someday; how soon depends on how far back from the shore the houses were built and the rate of erosion. North Carolina requires a setback of 30 times the annual erosion rate.

Move back

When existing shorefront homes are threatened by erosion, the costs and benefits of moving the structure back from the shore should be weighed along with the other alternatives. Depending on the nature of the problem, a moveback can compare favorably to other alternatives and prove to be economically, environmentally, and aesthetically better in the long run. The advantages and disadvantages of a moveback are similar to the construction setback; however, the moveback may have a high initial monetary cost. The cost of this solution is determined in part by the type of construction of the building to be moved, something that should be considered when building or buying a house.

Table 3.1 Common erosion-abatement techniques

Construction setback and moveback	Natural vegetation	Beach nourishment	Bulkheads or seawalls	Revetments	Groins
ADVANTAGES					
Reduces threat of destruction	Dampens waves and binds soil	Beach suitable for use	Shields the land from wave attack and erosion	Shields the land from erosion and reduces wave reflection	Builds beach on updrift side
Allows natural shore processes to operate	Reduces soil creep and rain wash	Does not have a negative impact on downdrift shoreline	Low maintenance costs	Individual units allow settlement and replacement	
No impact on marine biota	No impact on marine biota			Very durable	
DISADVANTAGES					
Does not halt erosion	Easily damaged by wave action	Does not halt erosion	Limits access and recreational use of beach and scenic value	Limits recreational use of beach and scenic value	Downdrift shoreline may erode
Area must be available for relocation	May have to be frequently replanted	Has to be periodically renourished	May cause loss of adjacent beach	May cause loss of adjacent beach	Beach nourishment may be needed in low longshore transport areas
May reduce views	Limits access and recreational use of beach	Affects biota	Affects biota	Affects biota	Affects biota

50 Shoreline engineering

Natural solutions

Salt-marsh grasses are found along many miles of the Chesapeake Bay shoreline. Where healthy wide strips of salt marsh fringe the shoreline, serious erosion is usually less of a problem (see chapter 2). In fact, since marsh and nearshore marine grasses generally reduce water currents and dampen wave energy, sediment will be deposited. This may build the shoreline up and out.

In addition to salt-marsh grasses that live in intertidal areas, subtidal sea grasses reduce shoreline erosion by mitigating wave energy. Unfortunately, many of the beds of marine grasses that used to cover large areas of Chesapeake Bay have disappeared. This extensive loss of the sea grasses on the Bay floor may have contributed to the erosion problem of some stretches of Bay shorelines.

In both Maryland and Virginia, planting of marsh grasses to reduce shoreline erosion problems has met with mixed success. Planting marsh grasses appears to work well in areas of moderate or low wave energy (fig. 3.3). However, in areas where wave energy is higher (shorelines facing a large amount of open water), marine grasses are easily damaged by storms. A number of studies have been done on the environmental impact of planting grasses as erosion buffers (reference 112, appendix C). Citizens interested in trying this relatively low-cost method of erosion reduction should review these reports and get in touch with the appropriate state agencies (listed in appendix B).

A major problem in introducing either sea grasses or marsh

3.3 Marsh grasses (*Spartina alterniflora* and *Spartina patens*) planted in a low-wave energy cove along the Choptank River are stabilizing this shoreline. The nearshore was graded to reduce the slope prior to planting.

grasses is waves. Stretches of shoreline exposed to large open areas of water may be subjected to storm waves of sufficient strength to tear out newly planted and even well-established grasses. Frequent replanting may be necessary. In addition, it may be necessary to grade the shoreline in order to produce a more gently sloping surface for grass planting.

Dunes, whether natural or man-induced, offer protection from storm waves and furnish a supply of sand to the beaches. This can slightly reduce the rate of erosion. Dunes are a common feature of certain shoreline segments in southern Chesapeake Bay and along the adjacent Atlantic Ocean, but are rare in the middle and upper Bay. The largest dune fields border portions of the open-ocean coast (fig. 3.4). Within Chesapeake Bay they most commonly occur on the sand spits that extend out from river mouths on Virginia's Eastern Shore and along the open Bay shoreline of Accomack, Mathews, and York counties.

Dunes can be built up by the use of sand fences and can be maintained by the use of vegetation and walk-over structures so as to prevent destruction of vegetation that holds sand in place. A sand fence is a fence, usually made of wood, set up in the dunes to trap blown sand. Eventually much of the fence is buried, helping to create dunes.

Artificial grasses or seaweeds have been used to mimic the effects of natural grasses. These devices are usually designed to dampen wave or tidal currents, thus reducing erosion and encouraging the deposition of sediment. This approach has been utilized on open-ocean, high-energy coasts such as the Outer Banks, North Carolina, and the New Jersey coast, as well as the quieter shorelines around the Great Lakes. Although short-term successes have been claimed, studies by the U.S. Army Corps of Engineers and North Carolina Sea Grant Program (reference 120, appendix C) indicate limited or little long-term impact.

3.4 (A) Extensive dunes located on Assateague Island, Virginia. The dunes are a major storage site for windblown sand which is somewhat stabilized by vegetation. (B) Sand fences have been installed in the dunes at Assateague Island to help "catch" sand. The sand fences along with natural grass planting are part of an effort to enlarge and stabilize the dune fields.

52 Shoreline engineering

3.5 Sand is frequently trucked to Virginia Beach, above left, and distributed with bulldozers in an effort to maintain the beach, below left. Sand also has been pumped onto the beach, above.

Beach nourishment: a "soft" solution

Replacing sand on a beach by an artificial means is known as *beach nourishment* or *replenishment*. Methods of putting on sand range from dump-trucking, the least expensive (if a source of sand is nearby), to pumping up sand from the adjacent sea floor (fig. 3.5). A big advantage of beach nourishment is that the recreational beach is maintained.

Nourishment is a fairly costly process. On open-ocean beaches, $1,000,000 per mile is a more-or-less minimal figure. In Chesapeake Bay expenditures of large sums of money on beach nourishment often are not economically justifiable. Also, since nourishment is, at best, a temporary solution, the costs must be borne again and again.

Beach nourishment, for all of the above reasons, is not a widespread approach to shoreline stabilization in the Bay. In the future it will probably be used mainly to build up small swimming beaches. Exceptions to this occur in the lower Bay near the entrance. A major beach-replenishment project was carried out in 1984 by pumping sand from Little Creek to City Beach at East Ocean View. Sand also has been trucked into Gloucester Point, Virginia, to widen the town beach. A total of 11,000 cubic yards of sand was used to replenish 1,300 feet of shoreline. Likewise, beach nourishment is frequently used on the open-ocean shore of Virginia Beach.

Hard solutions

Shore-perpendicular structures (groins and jetties). Walls of stone, steel, wood, or plastic bags extending perpendicular to the shoreline are called *groins*. Groins are supposed to trap sand, to build up a beach, and protect the fastland (mainland or land immediately adjacent to the beach).

Like groins, *jetties* extend perpendicular to the beach, but jetties are located at harbor entrances and are usually larger structures. Jetties are intended to make navigation safer by preventing sand from flowing into inlets.

In order for groins to succeed, the beach must have a good supply of sand that is moving or being pushed by waves (longshore transport) in one preferential direction along a beach. If the longshore transport is low, sand has to be brought in from elsewhere and dumped on the beach. In areas where the longshore transport of sand is high, the flowing sand builds the beach out on the "updrift" side of the groin (fig. 1.11). The problem is that the "downdrift" side becomes sand starved and erosion is likely to increase. Thus, one property owner's boom may be a neighboring property owner's bust. When one groin is emplaced in a reach of shoreline, several more often go up in self-defense. A whole series of *groins* is called a *groin field*. An example of a groin field occurs on the beach on Willoughby Spit in Norfolk, Virginia.

Groins come in a wide variety of designs (Fig. 3.6 and 3.7). They can be high, low, long, or short. In some cases they are made permeable on purpose so some sand will flow downdrift. In other cases, age and lack of maintenance causes groins to leak sand downdrift even though that wasn't originally intended.

Field experience in the Chesapeake Bay region has shown that "*low-profile*" groins are to be preferred. A low-profile groin is highest near the juncture with the fastland and tapers to a few inches off the bottom at its outboard end. In most cases the fastland end is constructed one to three feet above the existing beach level. Low-profile groins have the advantage of reducing sand starvation and subsequent erosion of the downdrift beach; the low-profile cross section allows sand to bypass the groin once the groin is approximately half full.

54 Shoreline engineering

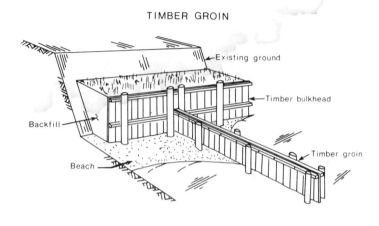

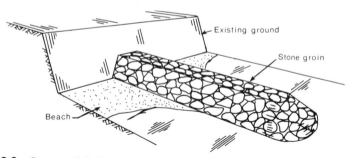

3.6 Stone and timber groins. Modified from *Shore Erosion Control* (reference 74, appendix C).

The best way to put in a successful groin or groin field (especially in Chesapeake Bay where the longshore transport of sand is generally low) is to nourish the beach with sand after the groins are constructed. Nourished beaches with groins tend to last far longer than nourished beaches without groins, but nourishment adds considerably to the cost.

Groins are often not the most successful way to stop shoreline erosion in Chesapeake Bay because of the limited longshore sand supply. In many areas there simply is not much sand flowing along the beach to be trapped by groins. For example, a shoreline in lower Gloucester County, Virginia, has groins of many different designs, but they have had limited success. The beach sand layer along this shoreline is very thin (on the order of six inches), so there basically is no sand available to trap. In Northampton County, Virginia, groins built out of railroad ties sunk vertically in the sand have also met with limited success. A gap of five feet or so was left in each groin in order to allow beach walkers access past the groins, and these gaps decreased the groins' ability to trap and hold sand. Gapping has also caused a groin field to fail in Accomack County, Virginia, but here nature produced the gaps. Shoreline erosion eventually flanked the landward ends of the groins leaving them unattached to the land. Betterton Beach in Kent County, Maryland has had better luck. A solid pier, 250 feet long, acts as a groin and has successfully held a pocket beach that originally was trucked in.

Jetties are responsible for some of the most significant human-induced erosion problems. A dramatic example of this occurred

Shoreline engineering 55

3.7 (A) Large stone groin at Ocean City, Maryland. (B) Small stone groins built to maintain a beach along the Choptank River, Maryland. Note that the groins are being used along with timber bulkheading. (C) and (D) Timber groins at Cape Charles City, Virginia. Note the buildup of sand on the updrift side of the groins. (E) Groin field in Virginia.

56 Shoreline engineering

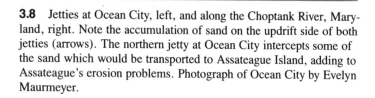

3.8 Jetties at Ocean City, left, and along the Choptank River, Maryland, right. Note the accumulation of sand on the updrift side of both jetties (arrows). The northern jetty at Ocean City intercepts some of the sand which would be transported to Assateague Island, adding to Assateague's erosion problems. Photograph of Ocean City by Evelyn Maurmeyer.

after the August 1933 hurricane opened an inlet through Fenwick and Assateague Islands, Maryland. Following the opening of this inlet, jetty construction was initiated on the north side. This approximately 1,500-foot-long jetty was completed in 1934, and an even-larger jetty was finished in 1935 on the south side. Since 1935 periodic construction has increased the lengths and heights of the jetties (fig. 3.8). The opening of Ocean City Inlet and the construction of these jetties significantly reduced the net southerly longshore flow of sand to Assateague Island. The result has been a buildup of sand updrift of the inlet near the jetty, and dramatic erosion (on the order of 30 feet per year) downdrift of the inlet at Assateague Island. The lesson is: Don't live downdrift from a jetty if you can avoid it.

If you do live downdrift from such a jetty, you can reduce the erosion problem by moving sand artificially past the jetties (e.g., dredging, hauling). The owner of a marina in Lancaster County, Virginia, has such a system to move sand from the updrift side to the downdrift side.

Shore-parallel structures (seawalls, bulkheads, revetments, and breakwaters). A large family of structures built on and parallel to the shoreline are seawalls, bulkheads, or revetments (riprap). *Seawalls* and *bulkheads* are vertical walls built primarily to hold back the forces of the sea, to keep the land from falling in, and to prevent erosion (figs. 3.9 and 3.10). *Revetments*, which have a similar purpose, are facings of stone, concrete, or other resistant material covering a slope (figs. 3.11 and 3.12). The actual distinction in definition between seawall and bulkhead is vague, and in many engineering publications they are used as interchangeable terms.

Shore-parallel walls are the most common shoreline-erosion–abatement structures in Chesapeake Bay. They are also probably the most successful, at least in terms of slowing shoreline retreat. As already pointed out, varieties abound and adjacent properties are often protected by different types of structures.

The property owner should consider several factors that are known to affect the success of shore-parallel structures. Obviously, the height of the structure is an important consideration. If a structure is overtopped by storm waves, flooding or erosion may occur behind the structure, which can cause early failure (figs. 3.13 and 3.14). The desirable height for a bulkhead or a revetment can be estimated by looking at nearby structures, by examining flood maps (which now include the effects of waves in the predicted flood levels), by reviewing records of historic water levels, or by using engineering calculations of wave height, water height, and swash runup height. It is generally accepted that a structure should be designed for the "100-year storm" (a flood level with a 1 percent probability of occurring in any one year).

The depth bulkheads penetrate into solid ground should be at least equal to the height of the structure. Solid ground begins at the more cohesive sediments (e.g., clay layer) under a beach and not at the surface of the sand. The assumption must be made that the sand is mobile and may be removed. This is particularly true during a storm. The most common cause of bulkhead failure is undercutting of the structure by wave scour until it collapses. This undercutting can leave the structure hanging and provide an

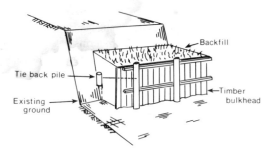
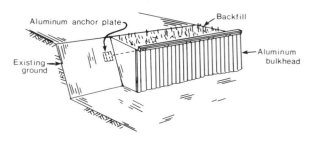
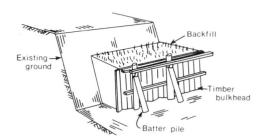
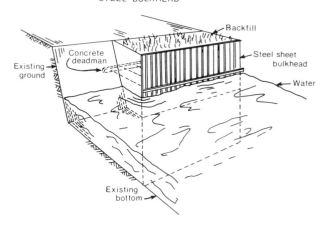

3.9 Common types of bulkheads found around Chesapeake Bay. All of the bulkheads shown here should extend below "hard ground" to a depth approximately equivalent to their heights (as shown for the steel bulkhead). Modified from *Shore Erosion Control* (reference 74, appendix C).

opening where soil at the rear of the structure will rapidly wash out. It is nearly impossible to repair the bottom of a structure or deepen a structure once it is built. Homeowners are often reluctant to pay for the underground portion of a structure because it is out of sight. Yet building a solid foundation is the best way to ensure that the structure will last even while the erosion of the shorefront continues. It is possible to place stones along the toe of an existing wall or bulkhead to diminish this undermining by waves.

It is now accepted that *"filter cloth"* should be used behind both bulkheads and revetments. This material is an engineered woven cloth that allows water to pass through freely, but prevents soil from washing out through any small opening or flaw in the face of the structure (figs. 3.15). Such washing out can be due to outflowing groundwater or the return of wave-splash water. Over time, washing out causes a cavity to develop, which often leads to the collapse of the structure or enlarging of the cavity. These cavities are common in older structures, and are very difficult to repair adequately.

Before filter cloth was in common use, the same effect was accomplished by building revetments first of small stone to serve as the filter, followed by courses of successively heavier stone. This design can still be functional, but filter cloth is both less expensive and more effective.

The larger stones forming the outer layers of a revetment must be carefully selected to be heavy enough to withstand the largest predicted waves. If they are not large enough, they can be hurled into your house or yard, and the structure will fail. Stone size will vary depending on the exposure of the location to large waves. The stones should be carefully placed or fitted on a revetment, rather than dumped. If stones are dumped, they may move around slightly during storms until they seat themselves, in the process leaving gaps where failure is likely.

The bulkhead or revetment is designed to be able to hold the fastland behind it in place. In an unstabilized bluff, groundwater is typically draining from the bluff. If an impermeable structure, such as a steel, wood, or cement bulkhead, is placed in front of the bluff, the groundwater will accumulate behind the wall. The additional water adds greatly to the forces pushing the wall out. This problem is easily solved by putting periodic *"weep holes"* that allow the water to drain, and prevent the buildup of pressure. Filter cloth will prevent soil from washing out of the weep holes. Countless older bulkheads in the Bay have failed for want of a simple weep hole.

In most settings, if a beach is desired in front of a wall, it most likely will have to be nourished from time to time, as the wall cuts off the immediate sand source for the beach. Nourishment can be expensive, but it serves to prolong the life of the wall as well as provide for recreation.

The design of both revetments and bulkheads should take into account possible *end-around effects* and wave or current scour at the base of the structures. The most well-designed structure, if flanked by unstabilized bluffs, will fail as the shorelines on the sides continue to erode and the ends of the structures begin to wash out. Experience in the Bay has shown that the ends of

A

C

B

D

E

G

F

3.10 (A) Timber bulkhead and (B) cement bulkhead located on the Choptank River, Maryland. The cement bulkhead has stone riprap at the base for protection against wave scour. Note the banks behind both bulkheads have been filled and graded. (C) Bulkhead at Virginia Beach, Virginia. (D) Wooden bulkhead at Gloucester Point, Virginia. Bulkheads have to be designed to withstand large waves, flooding, and ice effects. (E) Strong winds associated with Hurricane Gloria (1986) pound this bulkhead on the Choptank River, Maryland. Both flooding and large waves attacked this structure. (F) Ice on a bulkhead on the Choptank River, Maryland. (G) Tie-backs used to reinforce a bulkhead.

62 Shoreline engineering

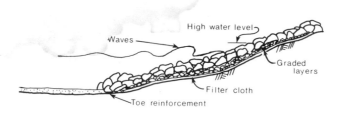
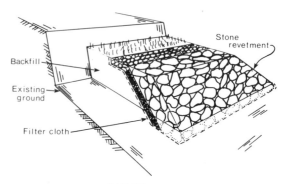
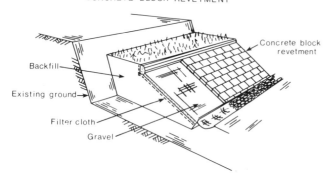
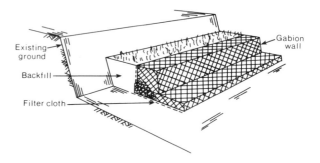

3.11 Various types of revetments found around Chesapeake Bay. Modified from U.S. Army Corps of Engineers (references 74 and 76, appendix C).

Shoreline engineering 63

3.12 (A) Stone revetment at Fort Story, Virginia. The preferred method for installing revetments is to fit the stones into place in order to prevent subsequent settling, which can lead to failure of the structure. (B) Similar to groins, revetments must be able to withstand storm waves, flooding, and ice effects. This revetment is located along the Choptank River, Maryland.

structures are susceptible to large washouts once exposed to wave action. The obvious solution to this problem is to build a single, continuous structure along any shore reach. If this is not possible, the ends of a structure should be carefully built into the fastland to minimize the end-around effect. Scour at the base of a structure can be minimized by ensuring that the structure penetrates far enough into the bottom (as previously discussed for bulkheads). A revetment at Fort Story, Virginia, made of 2,000 to 4,000 pound stones has undergone recent damage due to scour at the base of the rock pile. The loss of sediment at the toe of the structure is causing rocks to settle or fall into deeper water, leaving weakened zones in the revetment.

All structures built parallel to the shoreline will sooner or later fail. However, careful maintenance can increase the life of structures.

A good report on the cost and feasibility of various shoreline stabilization methods in Chesapeake Bay is published by the Maryland Department of Natural Resources (reference 89, appendix C). This report emphasizes case studies along the Maryland shore, but is highly applicable to the Virginia shore as well. Typical cost, in 1980 dollars, of stone revetments was $125 per foot of shoreline, $210 per foot for timber bulkheads, and $140 per foot for aluminum bulkheads. Protecting one's threatened property does not come cheap! Based on the study of 33 stabilization cases, this report favors the use of revetments rather than bulkheads for several reasons:

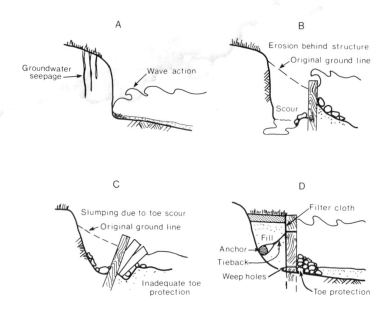

3.13 Influence of waves and overtopping on bulkheads. (A) Waves tend to cause scour at the base of a bank or a structure which causes the structure to rotate (C). (B) Overtopping causes fill material to be washed out, also leading to failure. (D) A bulkhead can be protected against these problems by making it penetrate an adequate depth into solid ground, installing stones as toe protection, anchoring the wall, using filter cloths, and replacing lost fill material. Weep holes are also needed to release groundwater. Modified from U.S. Army Corps of Engineers, 1981 (reference 76, appendix C).

1. The rocks used in revetments are less likely to deteriorate with time. If a revetment does fail in one spot, the entire structure is not likely to collapse, as is the case with bulkheads. Even if the structure fails, the rocks generally can be reused to repair the structure or can be piled along the toe of a structure to help dissipate wave energy.
2. Waves moving over stone blocks have part of their energy dissipated by the irregular surface. This means that there will be less reflection of wave energy seaward, as compared to a bulkhead, and less erosion of the beach. When a wave breaks on a revetment, the stone surface diminishes the amount of runup.
3. The irregular surface of a fitted stone structure provides a better habitat for organisms than a flat surface.

The Maryland Department of Natural Resources study also provides a number of case histories showing both successes and failures of structures. Several of these examples are given in the following paragraphs.

An example of a large structure that failed is a concrete bulkhead in Anne Arundel County, Maryland. After the five-foot-high wall failed, a house fell in. In York County, Virginia, some bulkheads made of unreinforced concrete blocks (a very poor design material for this purpose) quickly fell in during a minor storm. A timber bulkhead with a stone revetment in Queen Annes County, Maryland, was built too low. This has caused ponding of wave-splashed water behind the wall, which is eroding the backfill material.

3.14 Failed bulkheads in Chesapeake Bay.

Most bulkheads work better than the above examples. A bulkhead made up of vertical, stone-filled concrete well rings is working well near Baltimore. The high bank behind the bulkhead has been graded into a gentle slope and grassed. Gentle slopes behind bulkheads, as opposed to the preexisting natural steeper bluffs, aid greatly in holding back future shoreline retreat. Grading or terracing reduces the steepness of the slope, which helps attenuate wave energy as the wave runs up the slope (reference 74, appendix C). A 5:1 slope is recommended although 3:1 is often adequate (fig. 3.16). This method works best when used with other shore-protection strategies. But if a homeowner waits until his house is at the bluff edge before putting in a bulkhead, the grading option is no longer feasible.

In Northumberland County, Virginia, a reach with a very high historic erosion rate, erosion has been successfully reduced by a number of wooden bulkheads and groins, well anchored and built with pressure-treated lumber. Careful design and construction pays off!

Scientists Cliff, so-called because it is a world-famous fossil-collecting location for Miocene-age fossils and also because a number of scientists have homes nearby, is over 60 feet high in places and is composed of fairly durable sediment. The cliff is eroding at a rate of perhaps one foot per year through a combination of rain runoff and wave attack. Like all sea cliffs, it erodes particularly rapidly when accumulated material (*talus*) is removed from the base of the cliff by storms. Hence, to stop erosion, one must stop talus removal.

66 Shoreline engineering

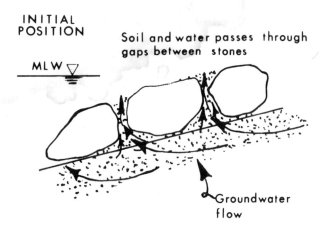

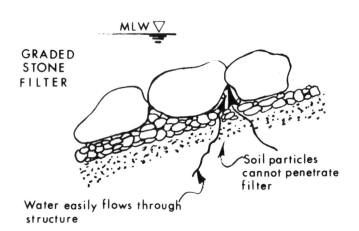

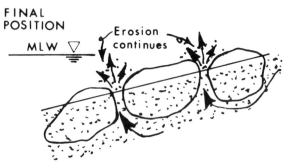

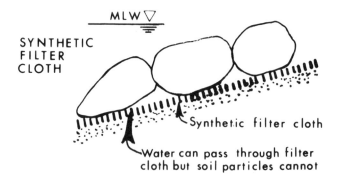

3.15 Influence of graded stone or filter cloths on revetments. Modified from U.S. Army Corps of Engineers, 1981 (reference 76, appendix C).

Shoreline engineering 67

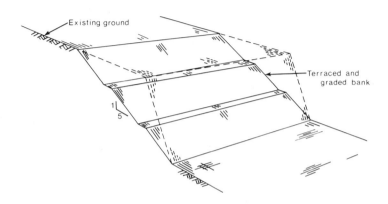

3.16 Grading a steep bank reduces slumping and erosion. It is useful to grade banks even when structures are emplaced. Modified from *Shore Erosion Control* (reference 74, appendix C).

Groins were emplaced at the base of Scientists Cliff 20 to 30 years ago. More recently *gabions* were put at the base of the cliff. Gabions are quite frequently used in Chesapeake Bay. They consist of "baskets," usually made of wire mesh, that are filled with rock or shell and then stacked on the beach like boulders in a revetment. In the case of Scientists Cliff the groins have not stopped waves from attacking the cliff base.

More successful gabions have been used elsewhere in Anne Arundel County, Maryland. In Accomack County, Virginia, oyster shells obtained gratis from local fishermen were used to fill gabions that have quite successfully halted local erosion.

Chesapeake Bay is the site of some rather unusual types of shoreline-erosion–halting schemes. The Craney Island dredge disposal site in Portsmouth is partially protected by highway slabs from the old James River Bridge. A commercial pier in Kiptopeke, Virginia, is protected by a row of sunken ships. The ships have reduced the wave energy striking the shoreline and the beach has built out as a consequence. At Fort Story, Virginia, the army has even thrown weapons of war into the breach by lining the shore with surplus amphibious vehicles.

Sunken ships are a special type of *offshore breakwater*. Breakwaters consist of structures placed offshore to reduce the energy of the waves striking the shoreline (fig. 3.17). A *"wave shadow"* is formed in the lee of the breakwater, and sand moving along the beach will halt in the wave shadow and build out the beach. The effect of breakwaters is not much different than the effect of groins because erosion may ensue in the downdrift direction. Breakwaters are also used to form quiet water for small-boat anchorages.

Offshore breakwaters can be made of almost anything ranging from massive concrete walls to old tires baled up and tied to the bottom (fig. 3.18). Some offshore breakwaters are floating structures. In Chesapeake Bay, offshore breakwaters, which are often costly to install and maintain, are at an early stage of development.

Three major breakwaters are located in Virginia at this writing. Two of these breakwaters are located on the Potomac River (Colo-

68 Shoreline engineering

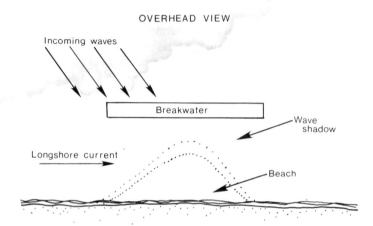

3.17 Overhead view of a breakwater. Modified from U.S. Army Corps of Engineers, 1981 (reference 76, appendix C).

nial Beach, Westmoreland County [fig. 3.19], and Aqua Po Beach, Stafford County) and are supposed to protect and maintain public beaches. The third breakwater, located along the James River (Drummond Field, James City County), was privately developed. At each site a major goal was the creation of a beach. Since there was no local source of sand, sand was hauled in and the beaches nourished. Although the cost of building breakwaters and nourishing the beaches is high, such measures protect relatively long stretches of shoreline. This ultimately yields a relatively low cost per linear foot.

A *sill* is a low breakwater that is backfilled to produce an elevated or *perched beach* (fig. 3.20). This process can provide both a broad buffer zone and a recreational beach (reference 76, appendix C). The sill must be able to hold the sand from being washed out, which may require filter cloth, etc. Also, the sill needs to be connected to the fastland to prevent end-around effects. Sills are often made of hard structures such as sheet piling, but can be made of less permanent material such as sand bags (fig. 3.21).

Unanchored structures. Another kind of shoreline-stabilization structure being developed and sold is referred to in this book as a "gravity structure." A *gravity structure* simply rests on the beach or bottom without a foundation extending below the surface; it is essentially unanchored. Examples include cement- or sand-filled bags or cement prisms. Gravity structures are designed to serve the same purpose as many of the more permanent hard structures, but at a lower initial cost. Similar to other coastal structures, gravity structures have potential problems that may reduce their useful life. Even after some short-term successes, the structures may fail and need to be replaced, driving up the "real" costs. Experience has shown that gravity structures are prone to the following problems:

1. Such structures are susceptible to damage from floating debris such as logs.
2. Alternate freezing and thawing during winter may take a severe toll on gravity structures, especially those of concrete, by breaking them up or generating weakening cracks.
3. The lack of foundation on a gravity structure may allow movement, which decreases its effectiveness.

CROSS SECTION
TYPES OF BREAKWATERS

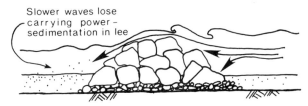

STONE RUBBLE BREAKWATER

Slower waves lose carrying power – sedimentation in lee

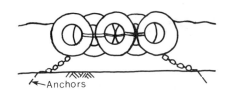

FLOATING TIRE BREAKWATER

Anchors

CONCRETE BAG BREAKWATER

Bags filled with lean concrete mix

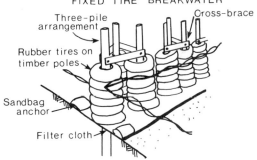

FIXED TIRE BREAKWATER

Three-pile arrangement
Cross-brace
Rubber tires on timber poles
Sandbag anchor
Filter cloth

3.18 Cross sections of common types of breakwaters. Modified from U.S. Army Corps of Engineers, 1981 (reference 76, appendix C).

3.19 Breakwaters located at Colonial Beach, Virginia.

Choosing shoreline-stabilization structures. When deciding what type of shoreline-stabilization structure to emplace, there are many factors to be considered, including which type of structure is best for the problem at hand and who to hire to install the structure. These questions are important because of the great expense of shoreline stabilization and the number of things that can lead to failure. There are a great many types of structures that have been designed to halt erosion, and new ones are frequently advertised. Many of these "new" designs are basically old designs with slightly different placements or different construction materials (e.g., a wall that zigs instead of zags, a wall of precast cement, and so on), and are not necessarily more effective than the old.

Because of the costs and importance of choosing the right structure, a landowner should do his homework. A good start is to pursue the suggestions outlined in this chapter and to obtain and read any available literature on methods or structures (see references 74, 76, 90, 120, etc., appendix C). It may also be wise to hire a coastal scientist to advise you. Do not be easily led by a salesperson who claims a "new" device is a major breakthrough or solution in the fight against erosion. Keep the following points in mind:

1. If a particular device does work and does cause sand to accumulate on a beach, the chances are good that a neighboring beach will suffer from a shortage of sand. After all, the sand you trap was going somewhere!
2. Many beaches in Chesapeake Bay are made up of a very

small volume of sand. Even if the device does not fail, there simply is a limited amount of sand to be trapped.
3. Devices that build up a beach under calm weather conditions may erode it during storms. Find out by asking the salesperson for local examples of where a particular system was installed and survived a season of storms. Reputable dealers will have demonstration projects and results that are verifiable.
4. Shoreline-erosion–prevention measures frequently do well for a year or two and then fail during "unexpected conditions."
5. Finally, does the seller/builder of the structure stand behind the product in terms of its ability to withstand storms and to protect the property behind it?

A philosophy of shoreline conservation

Methods of halting shoreline erosion are commonly called "shore-protection methods." There is much irony in this choice of words because shorelines do not need protection. Natural erosion is never a threat to a shoreline. It is only a threat to buildings and the other accoutrements of man. There is no shoreline-erosion problem until something useful to man is threatened or until man puts up a building by which to measure the erosion. As the shoreline retreats due to bluff erosion or whatever, it retains its same general appearance and its same attributes of size and shape. All that has happened is the shoreline has changed its position in space.

As a first step in preventing erosion problems in Chesapeake

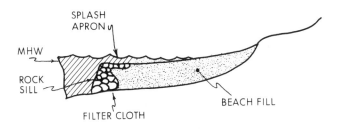

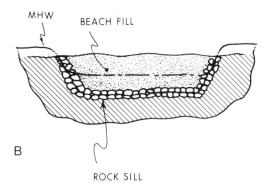

3.20 Perched beach construction: (A) cross-sectional view and (B) plan view. Modified after U.S. Army Corps of Engineers, 1981 (reference 76, appendix C).

72 Shoreline engineering

3.21 Sand-filled bags being used as shore protection in Chesapeake Bay.

property. Nevertheless, the necessity for shoreline stabilization frequently comes as a surprise to many new Chesapeake Bay shoreline dwellers. In any case, the compelling economic argument of the unreasonable cost of halting erosion on open Atlantic shores does not exist in the Bay.

Hundreds of miles of Chesapeake Bay shoreline now have no beach at all. Rocks and walls have replaced the beaches and swimming is either impossible or dangerous. The loss of the beach is often the price of saving thousands of homes and buildings that otherwise might have fallen into the Bay. But shoreline engineering has proven here in the Bay, as well as elsewhere, to be essentially an irreversible process. Once you start, you cannot stop. The first step toward shore protection represents a long-range commitment of effort and resources.

Bay we could build buildings farther back from the beach. As all Marylanders are aware, this has already been initiated with the "Critical Areas Criteria," which was passed into law in 1985 (discussed in chapter 6).

On open-ocean shorelines, most coastal states are trying to restrict building construction to areas behind some setback line. Such actions has been taken in the light of experience indicating that saving threatened buildings is very costly, and in the long run may end up destroying the recreational beach. Along the Chesapeake Bay shoreline, "permanent" erosion control can be emplaced at a cost within reach of most of those who buy shorefront

4 Selecting a site along the shores of Chesapeake Bay

Whether you are interested in purchasing a home or making an investment of some type close to the shoreline of Chesapeake Bay, you will find it prudent to maximize your knowledge of the region of your choice prior to making the investment! Although the sources are scattered, a great deal of information is available. Many of the publications, listed in appendix C, are available through various federal, state, and local agencies. For instance, the U.S. Army Corps of Engineers offices in Baltimore or Norfolk, the Maryland Department of Natural Resources, Coastal Resources Division, or the Virginia Marine Resources Commission are good sources. Addresses and other pertinent information for many of these agencies are found in appendix B.

In addition, you can learn a lot about the relative safety of a stretch of shore by visiting sites and making some basic, but very important observations. Chapter 2 described the various types of shorelines found in and around Chesapeake Bay and discussed the physical processes influencing each type. Many of the shore types described in chapter 2 indicate high rates of erosion and the potential of flooding. For instance, a lowland shoreline that faces a large, open expanse of water probably has a high erosion rate and a potential flooding problem. Construction of a home close to a shore of this type invites trouble. Expensive shore protection will no doubt be needed. On the other hand, a coastal area with a high elevation and a vegetated shore may well be a good bet.

A good start to evaluating a coastal site is to make a visit and check the area. Determine the type of shoreline and look for evidence of erosion or stability. Talking with local residents might give you information not available elsewhere.

When evaluating the suitability of a coastal site for your interests, consider the following list of characteristics:

1. Is the elevation of the site above the anticipated storm-surge level?
2. Is the site in an area of shoreline growth (accretion) or low shoreline erosion? Evidence of an eroding shoreline includes: (a) a bluff or scarp at the back of the beach; (b) stumps or peat exposed on the beach; (c) slumped features such as downed trees, bluff debris (talus); or (d) shoreline-stabilization structures such as seawalls, groins, or beach-replenishment projects.
3. For bluffed shorelines, is the site well back from bluff edge, even where erosion rates are low?

4. Is the site in a vegetated area (salt marshes, sea grasses, etc.) that suggests stability?
5. Is the site adjacent to a small stretch of open water? If a large distance of open water (fetch) exists in front of the site, storm waves will be larger.
6. Is the site behind a natural protective barrier such as a line of sand dunes?
7. For barrier islands, spits, or barrier beaches, is the site well away from migrating inlets? Is the site also away from low, narrow portions of the island or spit where the formation of a new inlet is possible?
8. Is the site in an area of either no or little historic overwash?
9. Does the site drain water readily?
10. If hookup to municipal water supply is unavailable, is the groundwater supply adequate and uncontaminated? Is there room for proper spacing between water wells and septic systems?
11. Are the soil and elevation suitable for efficient septic-system operation?
12. Is the soil adequate for the emplacement of footings?
13. Are adjacent shoreline protection structures adequately spaced and of sound construction? Are these structures eventually going to cause your shoreline to erode?

Visiting the site prior to making any decision to invest a lot of money is extremely important—whether the decision is to buy property, build a structure, or to install shore protection. Equally important is doing some homework. This means locating information on the region such as erosion rates, flooding hazards, or bluff-failure problems. It may not hurt to bring in an outside specialist to help in your decision—someone who is not involved in selling you the property or contracting the work! A number of sources of information and agencies who can help are listed in appendixes B and C. In addition, some of the coastal hazards for the entire mainstem of Chesapeake Bay and portions of its tributaries are presented in this chapter.

Shoreline descriptions and site-analysis maps

To provide general information on erosion problems and potential flooding hazards, we have divided the shoreline of Chesapeake Bay into 36 segments and have made site-analysis maps for each segment. Each site-analysis map shows the average erosion rate for a section of shoreline and the area of land that would be inundated by a 100-year flood. The table included with each site-analysis map gives the erosion rate. In addition, the tables that accompany the Virginia site-analysis maps provide general information on coastal geomorphology (shoreline types). The maps are grouped for convenience by geographic region and county. A general description of the area and some historical information on coastal hazards are also provided.

The erosion information presented on the site-analysis maps and tables was obtained from two sources. For Virginia's portion of Chesapeake Bay, changes in shoreline positions have been determined by the Virginia Institute of Marine Sciences (VIMS) and published in a series of documents entitled *Shoreline Situation Reports*. The VIMS reports use historical maps and aerial photographs to record changes in shoreline positions between circa 1850 and the present. They then compute an average shoreline erosion (or accretion) rate. There is a separate *Shoreline Situation Report* for each county in Virginia. More information concerning these reports is given in appendix C (reference 102). The original *Shoreline Situation Reports* provide more detail than the site-analysis maps given here, each of which covers a large stretch of shoreline. If a region is of interest after viewing the site-analysis map provided in this chapter, then the appropriate *Shoreline Situation Report* should be consulted.

For Maryland, the information concerning shoreline changes was obtained from a series of publications prepared by the Maryland Geological Survey (located in Baltimore, Maryland) entitled *Historical Shorelines and Erosion Rates* (appendix C, reference 103). These reports show changes in shoreline positions that have occurred since the region was mapped (as early as 1841) by the U.S. Coast and Geodetic Survey and more recent mappings (since circa 1940). These original reports contain much more detail than do the site-analysis maps presented in this volume. It would be wise to review the more detailed maps before making a final decision on land purchase or development.

In addition to shoreline stability (erosion), the prospective coastal investor must consider flooding hazards. In the Chesapeake Bay region the potential damage from floods is difficult to assess. In some instances, a storm surge may mean nothing more than an inconvenient inundation of water around one's property. On the other hand, if flooding occurs during strong winds when waves are being whipped up and strong currents driven, flooding may lead to serious damage and the loss of valuable property or worse! In Chesapeake Bay, the potential risks for a variety of types of floods have been determined by the U.S. Federal Emergency Management Agency (FEMA) and published on Flood Insurance Rate Maps (FIRMS) or Flood Hazard Boundary Maps (FHBMS). The FHBMS are not as accurate as the FIRMS, but they approximate the flood levels reasonably well. These maps, which are available through FEMA (see chapter 6 and appendix B) or through most local insurance companies, typically show the landward limit or the elevation of the 100-year flood level (the elevation that flood waters have a 1 percent chance of reaching in any given year). The general position of this elevation is presented on the site-analysis maps. As is the case for the shoreline-erosion information, the site-analysis maps generalize the 100-year flood lines. Before purchasing coastal property, the prospective buyer should consult a local insurance broker and the latest FIRM or FHBM for that area to determine if the site is within the 100-year flood zone. A relatively small investment of time and money could save a considerable amount of money and frustration in the future.

Using the site-analysis maps

The site-analysis maps provided in this chapter give preliminary information on long-term trends of shoreline erosion and flooding problems in Chesapeake Bay. If you are interested in a shoreline segment, first find the general location of the area on the map of Chesapeake Bay (fig. 4.1). This location map is subdivided into 36 segments and provides the number of the site-analysis map for each segment. It also shows important geographic locations. Next, proceed to the appropriate map in the following sections. The table of contents gives the exact page number. The site-analysis maps give the rate of shoreline change (erosion or accretion) as slight, moderate, high, severe, or accretion (table 4.1). Accretion refers to a seaward movement of the shore. Because two different sources were used to produce the maps, the actual rate of change for each of the categories differs for Virginia and Maryland. These groupings are based on the original sources discussed in the preceding section. It should be noted that during the production of these maps, stretches of shoreline that showed limited variability were grouped together for the sake of simplicity. Therefore, a particular shoreline area within a segment may have a different erosion rate than is shown. For instance, a segment may have both moderate and severe erosion rates, depending on the exact location. The site-analysis map normally will show only the highest erosion rates. More complete information is given in the table that accompanies each site-analysis map.

The areas likely to be inundated by tidal waters during 100-year floods are designated by a hatch pattern on the site-analysis maps. If the section of shoreline of interest is located along a severely eroding coast that is well within the 100-year flood line, exercise caution when purchasing the property. The area will likely need to be stabilized with some sort of expensive shoreline structure and may be flooded periodically. If a coastal area has shown only slight erosion and is well above the 100-year flood limit, it may be a good bet. As discussed in chapter 2, exercise caution if the shore is composed of a bluff that is subject to slumping.

Table 4.1 Classification of erosion rates used on site-analysis maps for Maryland and Virginia's Chesapeake Bay shoreline

Category	Average erosion rate per year (feet)	
	Virginia	Maryland
Slight	0–1	0–2
Moderate	1–3	2–4
High	3–6	4–8
Severe	more than 6	more than 8

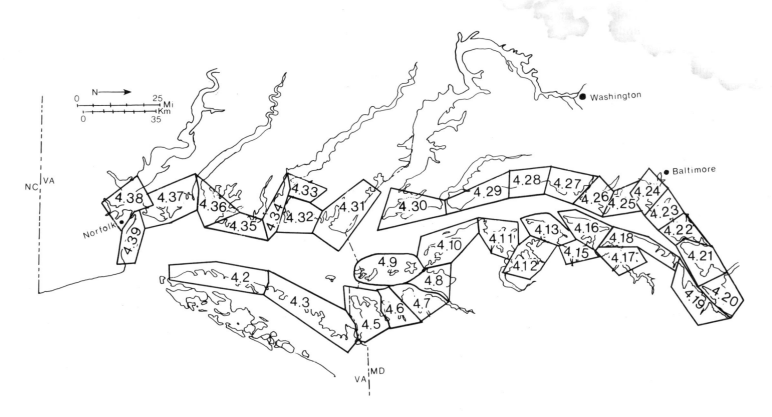

4.1 Map showing boundaries of site-analysis maps.

Eastern Shore, Virginia: Northampton and Accomack counties

The Eastern Shore of Chesapeake Bay in Northampton and Accomack counties consists of high sandy bluffs to the south (fig. 4.2, table 4.2), and an extensive low-lying region dominated by salt marshes to the north (fig. 4.3, table 4.3). The change in shore types closely matches the border of Northampton and Accomack counties.

The Bay-facing shores often are backed by moderately high bluffs (greater than 10 feet) (fig. 4.4), which elevate most buildings above flood level. These bluffs are incised by tidal creeks every few miles. As erosion takes place, a large amount of sediment slumps onto the beach and is quickly moved away by wave action. While much of this sand is carried offshore to form extensive sandbars, some is carried alongshore to the south, where it forms the sand spits that cross creek mouths. An example is found at the southern end of Church Neck and Savage Neck. These low spits are unstable because they are reworked by waves and tidal currents. Sand is also trapped by the Kiptopeke Pier and other structures to the south.

Almost the entire Bay-facing shoreline along Northampton County is experiencing erosion. Although the average retreat of the shore is over 2 feet per year, an extreme case is the Tankards Beach area of Savage Neck, where erosion is up to 20 feet per year. Erosion rates of 6 feet per year occur at Silver Beach on Occohannock Neck.

The entire coastal region of Accomack County is very low in elevation, with dozens of square miles of land lower than the 100-year flood level. As a consequence, the flooding hazard is high throughout the coastal region. Where coastal bluffs do exist, they tend to be low, and are muddy rather than sandy. These eroding bluffs do not feed much sand onto the beaches, which causes adjacent beaches to be thin and narrow.

The embayments that cut into the shoreline are protected from the large waves that cause rapid shore erosion. In these areas, salt marshes flourish. The combination of marsh and small waves leads to more stable shorelines that often show little or no erosion. However, these areas often experience flooding during storms. Areas of coastal flooding without wave impact may be suitable for elevated and flood-proofed buildings, whereas areas exposed to larger waves may not.

Lower Eastern Shore, Maryland: Somerset, Wicomico, and lower Dorchester counties

The lower Eastern Shore region of Maryland is characterized by vast marshes that are unparalleled both in their beauty and their value as wildlife habitats. Unfortunately, a characteristic of marshes is that they have very low elevations and are frequently flooded. Much of the coastal area of Somerset is composed of marsh with intermittent sandy pocket beaches (figs. 4.5 and 4.6, table 4.4). The main population center in coastal Somerset is the town of Crisfield, which is located in the more protected confines of Little Annemessix River. Although erosion here is less severe than along the shore facing the more open waters of Chesapeake Bay, the low elevation of the area allows periodic flooding. Farther to the north in upper Somerset County and into Wicomico County and Dorchester County, the coastal area changes somewhat and much of the shoreline terminates in low sandy banks or lowlands though marshes still abound (figs. 4.7 and 4.8, table 4.5). From Deal Island to Dames Quarter the shoreline property owners have used numerous groins and bulkheads to mitigate erosion. Although lowlands have a higher elevation than the surrounding marshlands, flooding is still a problem. Much of this area is below the 100-year flood elevation as shown on FIRMs. Along the banks of the Nanticoke River from the town of Nanticoke to past Bivalve, the same story holds true. This area has somewhat less of a tidal flooding problem.

Similar to the mainland, the chain of offshore islands including Bloodsworth, South Marsh, Smith, and Tangiers islands (fig. 4.9, table 4.6) are composed almost entirely of marsh, and are affected by high rates of erosion and frequent flooding. The communities of Ewell, Rhodes Point, and Tylerton, which have an average elevation of approximately 2 feet and a maximum elevation of approximately 5 feet above mean sea level are inundated during extremely high tides, and are occasionally forced to evacuate during storm tides. During the storm on August 23, 1933 ("the August storm"), a surge of approximately 7.5 feet above mean sea level caused extensive flooding, which inundated a number of homes. This same story was repeated during Hurricane Hazel in 1954 and again in 1962. These islands, first described by Captain John Smith in 1608, have been significantly reduced in size by erosion.

Despite the problems of rapid erosion and potential flooding, these islands are not without development pressures. According to an article published in the *Baltimore Sun* in May 1986, a zoning request was submitted to build 100 townhouses, an 80-slip marina and a 2,000-foot airstrip on Smith Island. An interesting debate over this potential development rose with those in opposition arguing that a development of this size would damage the island's unique way of life. At this writing Smith Island has a year-round population of approximately 600 residents, a number that has been declining over the years.

Toward the north in the Hooper and Taylors islands areas (fig. 4.10, table 4.7), elevations are somewhat higher, but flooding and shoreline erosion remain major problems. On the Bay-facing

80 Selecting a site

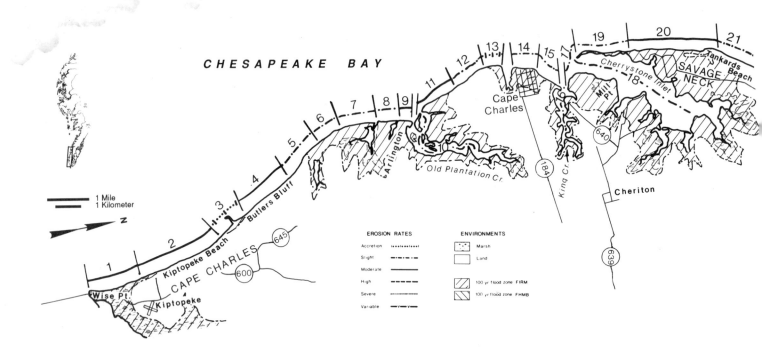

4.2 Site analysis: Northhampton County, Virginia. See table 4.2 for additional information.

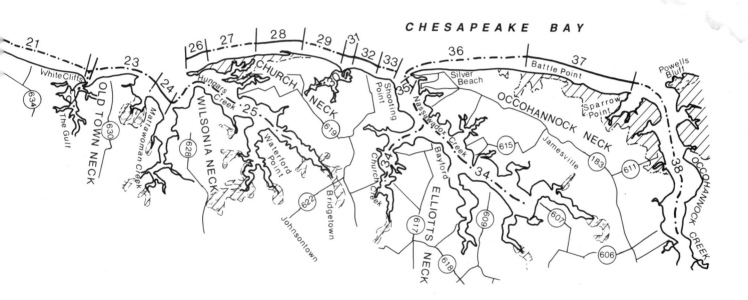

Table 4.2 Shoreline types and erosion rates: Northampton County, Virginia

Segment number	Shoreline type	Erosion rate	Segment number	Shoreline type	Erosion rate
1	Beach with dune	Moderate	20	Beach with dune	Moderate
2	Beach with high bluff	Moderate	21	Beach with high bluff	Slight
3	Beach with high bluff	Accretion	22	Fringe marsh	Slight
4	Beach with high bluff	Moderate	23	Beach with high bluff	Slight
5	Beach with dune	Slight	24	Beach with low bluff	Moderate
6	Beach with high bluff	Slight	25	Fringe marsh	Slight
7	Beach with dune	Slight	26	Spit with dune	Moderate
8	Beach with low bluff	Slight	27	Beach with dune	Slight
9	Beach with dune	Moderate	28	Beach with low bluff	Moderate
10	Fringe marsh	Slight	29	Beach with low bluff	Slight
11	Beach with low bluff	Moderate	30	Fringe marsh	Slight
12	Beach with dune	Slight	31	Spit with dune	Moderate
13	Man-modified	Accretion	32	Beach with high bluff	Moderate
14	Man-modified	Slight	33	Beach with high bluff	Slight
15	Beach with dune	Slight	34	Fringe marsh	Slight
16	Fringe marsh	Slight	35	Spit with dune	Slight
17	Beach with low bluff	Slight	36	Beach with high bluff	Slight
18	Fringe marsh	Slight	37	Beach with low bluff	Moderate
19	Spit with dune	Slight	38	Fringe marsh	Slight

shore, the shoreline is receding at an average of more than 8 feet per year in many areas. Both Taylors and Hooper islands have been severely damaged by storm activity. During the 1933 August storm, which caused extensive damage to Smith Island, Taylors and Hooper Islands were also heavily damaged. According to the *Banner*, the newspaper of Cambridge, Maryland, the two main bridges (Narrows Ferry and Fishing Creek) at Hooper Island were swept away in the storm: "The draw tender house which stood on a shell pile just off the upper Island end of the Narrows Ferry Bridge was last seen floating more than a half a mile away from its customary place and the tender James Riggin and his assistant G. Cleveland Riggin had not been located yesterday morning." One body was found later in the Bay. During this devastating storm numerous homes, boats, and businesses were destroyed. The *Banner* reported an eyewitness account of the final hours of the White and Nelson crab house on Hooper Island: "At the time that the crabbing house was being beat to pieces by the heavy seas and winds, there were about 60 men, women and children working in the factory and when it became apparent that the building was becoming unsafe, they were placed in a large boat and started for the shore: this boat struck a sunken piling and started to sink. The people were transferred to another boat and this one sank just before the shore was reached and those in it jumped overboard and waded ashore."

By bringing back an unusually harsh event, this account illustrates the potential for major flooding in the region.

Middle Eastern Shore, Maryland: upper Dorchester, Talbot, and Queen Annes counties

South of the entrance to the Choptank River in the Trippe Bay area (fig. 4.11, table 4.8), the open, Bay-facing shoreline suffers from the same flooding and erosion problems as does its neighbors further to the south on Taylors and Hooper islands. This shoreline, which is characterized by low eroding banks cutting into the lowland topography, is receding at rates of over 8 feet per year in some places. In sharp contrast are the more stable shore areas within the protected confines of the Little Choptank River. Here erosion rates are low, but flooding remains a problem. Although the shoreline erosion problem is not as severe as it is on the bayward shores, shoreline recession may accelerate in the outer areas of the Little Choptank as James Island and Hills Point continue to disappear.

Inside the Choptank River on both the southern and northern shores (fig. 4.12, table 4.9; and fig. 4.13, table 4.10), the rates of shoreline erosion are highly variable. In some of the more exposed areas like Cook Point or Benoni Point, shoreline erosion is high. In more protected embayments and tributaries, the shore is much more stable, suffering only slight erosion. Again flooding can be a nuisance for much of the land around the Choptank River. Almost all of the shoreline of Cambridge has protective structures (fig. 4.14).

A much different condition exists along the bayward shores. Erosion rates on the bayward side of Tilghman Island (fig. 4.13, table 4.10) are frequently high to severe. Historical erosion rates at

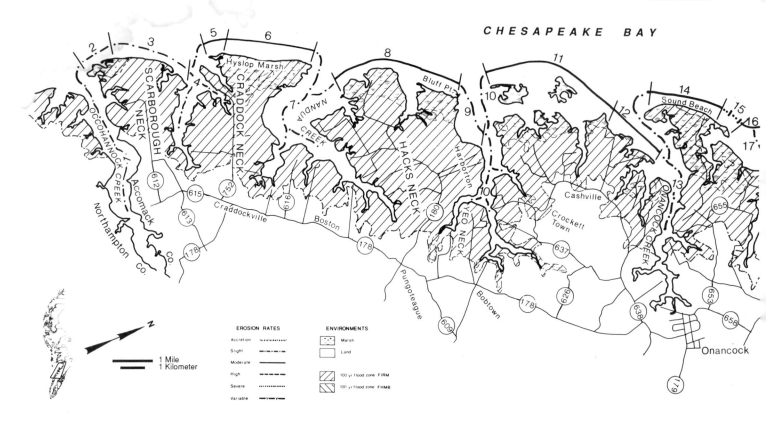

4.3 Site analysis: Accomack County, Virginia. See table 4.3 for additional information.

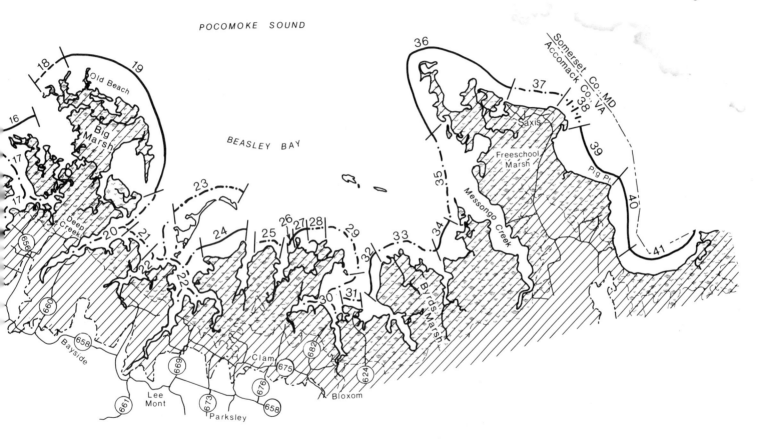

Table 4.3 Shoreline types and erosion rates: Accomack County, Virginia

Segment number	Shoreline type	Erosion rate	Segment number	Shoreline type	Erosion rate
1	Fringe marsh	Slight	22	Fringe marsh	Slight
2	Spit with dune	Slight	23	Beach with low bluff	Slight
3	Beach with dune	Slight	24	Beach with low bluff	Moderate
4	Fringe marsh	Slight	25	Marsh margin with sandy barrier	Slight
5	Spit with dune	Moderate	26	Beach with low bluff	Moderate
6	Marsh margin with sandy barrier	Moderate	27	Marsh margin with sandy barrier	Moderate
7	Fringe marsh	Slight	28	Beach with low bluff	Slight
8	Marsh margin with sandy barrier	Moderate	29	Marsh margin with sandy barrier	Slight
9	Beach with low bluff	Slight	30	Fringe marsh	Slight
10	Fringe marsh	Slight	31	Beach with low bluff	Slight
11	Marsh margin with sandy barrier	Moderate	32	Marsh margin with sandy barrier	Moderate
12	Beach with low bluff	Moderate	33	Marsh margin with sandy barrier	Slight
13	Fringe marsh	Slight	34	Marsh margin with sandy barrier	Moderate
14	Marsh margin with sandy barrier	Moderate	35	Marsh margin with sandy barrier	Slight
15	Spit with dune	Accretion	36	Marsh margin with sandy barrier	Moderate
16	Marsh margin with sandy barrier	Moderate	37	Beach with low bluff	Slight
17	Fringe marsh	Slight	38	Spit with dune	Accretion
18	Marsh margin with sandy barrier	High	39	Marsh margin with sandy barrier	Moderate
19	Marsh margin with sandy barrier	Moderate	40	Beach with low bluff	Moderate
20	Fringe marsh	Slight	41	Marsh margin with sandy barrier	Moderate
21	Marsh margin with sandy barrier	Slight			

4.4 Eroding bluff in Northampton County, Virginia.

the southern end of Tilghman Island exceeded 20 feet per year for the period of 1847 to 1942. A stone revetment located along part of the southern end of Tilghman Island near Black Walnut Point has alleviated the severe rate in this area, although at a high cost.

Along the southern shore of the Chester River and in the Eastern Bay region (fig. 4.15, table 4.11), most of the shoreline is protected from direct wave attack, reducing erosion rates. However, shore-protection structures are still needed in many areas. Flooding potential is high close to the water. Both flooding and erosion problems are reduced in areas with higher elevation. On Kent Island, the bayward shore has a generally higher elevation, reducing flooding hazards (fig. 4.16, table 4.12).

Upper Eastern Shore, Maryland: upper Queen Annes, Kent, and Cecil counties

The Chester River, which forms the boundary between Queen Annes and Kent counties, has historically had low erosion rates, averaging less than 2 feet per year in most places (fig. 4.17, table 4.13). In fact, many stretches of the shore have accreted. This growth is not surprising as most of the river has very limited open water or fetch, minimizing wave force. Unfortunately, the advantage of the more stable shoreline is offset by the greater risk of flooding, especially along the Kent County shoreline. Nearly all of the northern shore of the Chester River is within the region that will be flooded by the 100-year flood. Conversely, some of the

88 Selecting a site

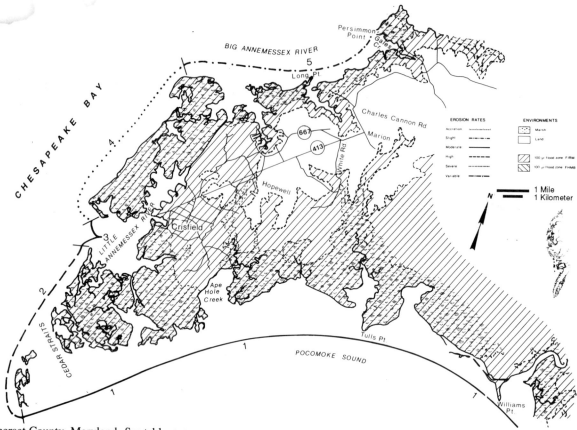

4.5 Site analysis: lower Somerset County, Maryland. See table 4.4 for additional information.

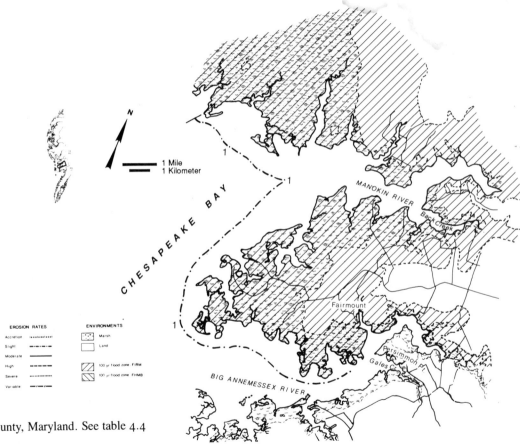

4.6 Site analysis: middle Somerset County, Maryland. See table 4.4 for additional information.

Table 4.4 Erosion rates: lower and middle Somerset County, Maryland

Segment number	Erosion rate
Lower Somerset County	
1	Primarily slight to moderate; isolated areas of severe erosion
2	Moderate to high
3	Primarily slight to moderate
4	Moderate to severe
5	Accretion to slight
Middle Somerset County	
1	Accretion to slight

southern shore of the Chester River has higher elevation, reducing flooding hazards.

In sharp contrast to the more protected areas of the Chester River, the Kent County shoreline facing into Chesapeake Bay is exposed to much greater wave forces and consequently is eroding at a higher rate (fig. 4.18, table 4.14). From Eastern Neck Island to Rock Hall, much of the coast consists of low rolling hills fronted by beaches, which are retreating at slight to moderate rates. North of Rock Hall, the land becomes more rugged and much of the coast consists of eroding bluffs. Along the stretch from Worten Point to Betterton, much of the shoreline consists of bluffs 10 to 50 feet in height (fig. 4.19, table 4.15).

Because of the generally higher elevation of the land from Swan Point to the Sassafras River, very little of the coastal region is below the 100-year–flood elevation. Consequently, flooding is not as great a potential problem as in many other areas. Still, the area has experienced relatively high water levels during storms. During the devastating August 1933 hurricane, the tides were 8.2 feet above mean sea level at nearby Baltimore. The predicted storm surge for the 100-year flood is approximately 9 feet at Tolchester Beach and 10 feet at Betterton (see chapter 1, table 1.1).

North of the Sassafras River to Town Point Neck, the shoreline is relatively stable compared to many other areas of Chesapeake Bay. The erosion rates here are predominantly slight and many areas have shown some accretion since the mid–nineteenth century. The exception to this occurs near Grove Neck, where the shoreline is undergoing moderate rates of erosion. In addition to the relatively low erosion rates, the flooding potential is minimal with mostly the low-lying marshes within the 100-year–flood area.

Elk Neck, the peninsula bounded by the Elk River on the east and the Susquehanna Flats on the west, is composed of rolling hills that frequently terminate in actively eroding bluffs. The rate of the coastal retreat in this area has been for the most part slight, and flooding hazards are limited to the small embayments interrupting the bluffs (fig. 4.20, table 4.16). The areas around the mouth of the Susquehanna River (from the Northeast River to Perryville) show similar stability with much of the area showing only slight erosion and being above the predicted 100-year flood elevation.

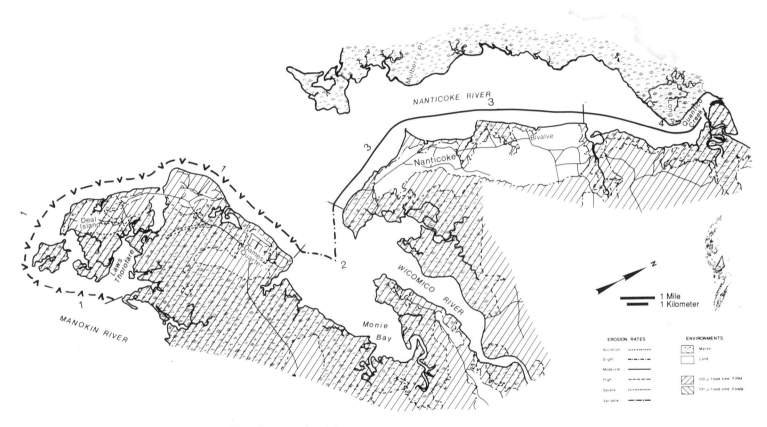

4.7 Site analysis: upper Somerset and Wicomico counties, Maryland. See table 4.5 for additional information.

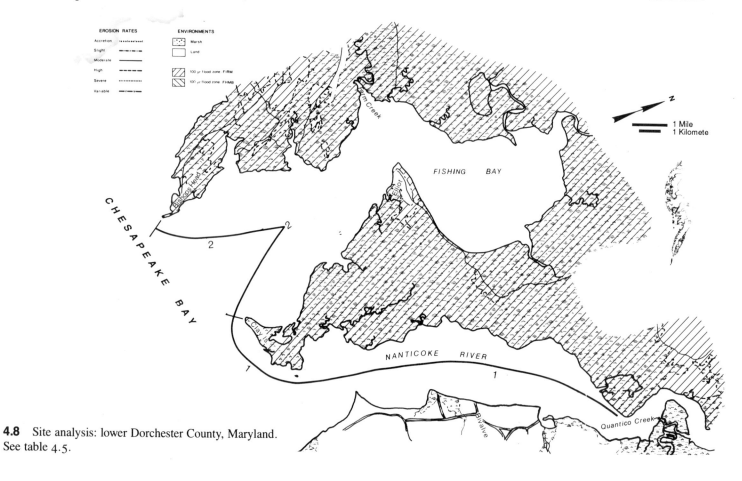

4.8 Site analysis: lower Dorchester County, Maryland. See table 4.5.

Table 4.5 Erosion rates: upper Somerset and Wicomico counties and lower Dorchester County, Maryland

Segment number	Erosion rate
1	Variable (slight to high)
2	Primarily accretion to slight; some moderate
3	Primarily slight to moderate
4	Slight to moderate; some accretion
Lower Dorchester County	
1	Slight to moderate; some accretion
2	Slight to moderate

Upper Western Shore, Maryland: Harford, Baltimore, and upper Anne Arundel counties

Much of the coastal area of upper Harford County is part of the Aberdeen Proving Ground and is consequently unavailable to the public for development. This region has numerous low, eroding banks that are periodically fronted by beaches and interrupted by marshes (fig. 4.21, table 4.17). Erosion rates range from slight to moderate except for the more rapidly eroding marsh areas toward Sandy Point. The most concentrated development occurs at Havre de Grace where much of the shore is protected by erosion-control structures. Despite the proximity to both the Susquehanna River and Chesapeake Bay, flooding hazards in Havre de Grace are essentially limited to the immediate shoreline area. Much of Havre de Grace stands higher than the 100-year–flood level.

The coastal area along the upper Gunpowder River is largely made up of low eroding banks, marsh, and occasional beaches (fig. 4.22, table 4.18). Most residential development is in the upper reaches of the Gunpowder River delta area (Harewood Park, Joppatowne). Here erosion rates are low, largely due to the lack of open water. The advantages of the more stable shore are offset somewhat in low-lying areas by the flooding problem. Areas above the predicted maximum storm-surge level are in relatively low-risk environments.

Between Gunpowder River and Back River (fig. 4.22, table 4.18; and fig. 4.23, table 4.19) erosion rates are for the most part very low due to the lack of appreciable open water. Little of the

94 Selecting a site

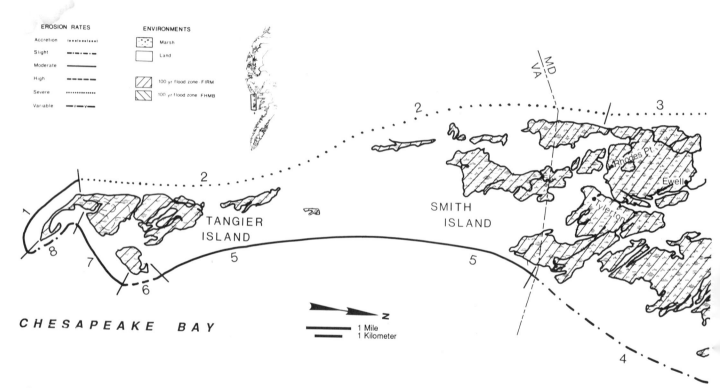

4.9 Site analysis: Bloodsworth, South Marsh, Smith, and Tangier islands. See table 4.6 for additional information.

Selecting a site 95

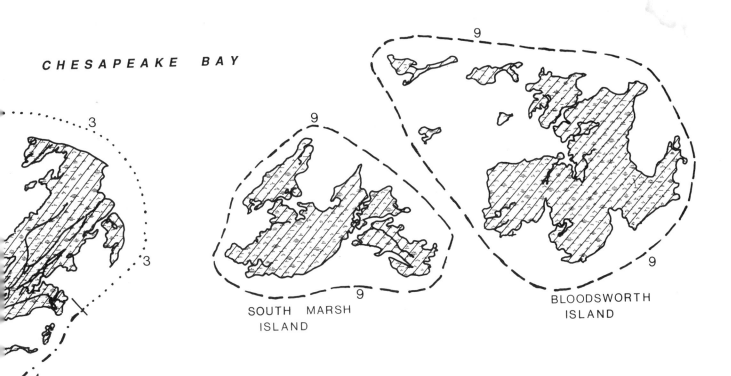

Table 4.6 Erosion rates: Bloodsworth, South Marsh, Smith, and Tangier islands, Maryland

Segment number	Erosion rate
1	Moderate
2	Severe
3	High to severe
4	Primarily slight; some moderate
5	Moderate
6	High
7	Moderate
8	Slight
9	No information available; probably moderate to high

developed area is considered to have critical flooding problems or is within the area that would be inundated during a 100-year flood (shown on FIRMS). The exception to this occurs in coastal areas with elevations less than approximately 10 feet.

The most extensively developed area in Maryland's Chesapeake Bay is found within the Baltimore Harbor (fig. 4.24, table 4.20). Toward the entrance of Baltimore Harbor, the shoreline is undergoing slight erosion. Within Baltimore Harbor, however, most of the region is heavily developed and the coastal areas are protected by numerous erosion-control structures.

South of the Patapsco River to the South River, much of the landscape is composed of gently rolling hills with heights up to or over 80 feet (fig. 4.25, table 4.21; and fig. 4.26, table 4.22). In the bayward exposures of the Gibson Island or Broad Neck areas, these hills are actively being eroding to form steep bluffs. In the more protected areas of Magothy River, erosion rates are lower. In parts of the South and Severn rivers, the shorelines are frequently composed of steep, eroding bluffs (fig. 4.27, table 4.23). In more protected areas where wave erosion is less, the bluffs may have gentler slopes and may be vegetated.

Rates of shoreline erosion are highly variable in this area depending on shore type and exposure to wave action. Extremely high rates of erosion have occurred along exposed headlands like Sandy Point or Hacket Point. On the other hand, the more protected embayments, where much of the development is located, have low erosion rates. The higher bluffs, although eroding, typically have low erosion rates. Caution must be exercised when developing near an eroding bluff. Although bluff retreat may be slower, they can fail catastrophically. See the discussion of bluff processes in chapter 2.

Flooding problems are largely confined to the areas immediately adjacent to the shoreline. The predicted 100-year–flood level for Annapolis is a little over 9 feet, although this surge level does not account for additional flooding due to wave activity. Much of the area south of the Severn River is above this level, but there is no guarantee that flood levels will not be higher.

Lower Western Shore, Maryland: lower Anne Arundel, Calvert, and Saint Marys counties

Much of lower Anne Arundel County's shoreline is composed of low eroding bluffs fronted by sandy beaches. Here erosion rates are highly variable ranging from slight to high, and some of the area is prone to tidal flooding (fig. 4.28, table 4.24). Toward the south in Calvert County, the shoreline is composed of large bluffs with elevations greater than 50 feet in many areas. Most of these bluffs are actively eroding at relatively low rates (fig. 4.29, table 4.25). Although these bluffs are exposed to open Chesapeake Bay, they experience lower rates of erosion because slumping has placed a great amount of debris at the base of the bluffs. Slope failure is partly related to oversteepening of the face because of active erosion at the base. When a bluff slumps, material falls to the base of the bluff and must be removed by wave action before the bluff becomes undermined. Slumping also exposes the spectacular sharks' teeth and other fossils found in areas such as Scientists Cliffs. Of course, other processes are involved in bluff slumping (see chapter 2).

Along this stretch of coast the risk of tidal flooding is limited to low-lying areas that locally dissect the higher bluffs. South of the Patuxent River in Saint Marys County the risk of tidal flooding increases (fig. 4.30, table 4.26). This is especially true south of Point No Point and into the Potomac River, where much of the area is below the 100-year–flood elevation.

98 Selecting a site

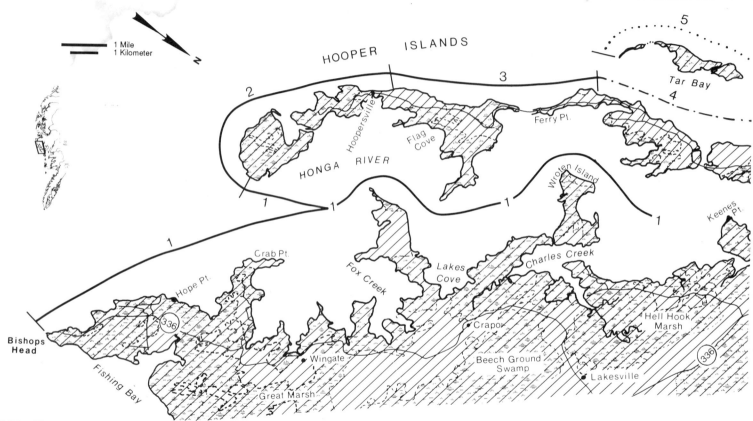

4.10 Site analysis: Dorchester County, Maryland (Taylors and Hooper islands). See table 4.7 for additional information.

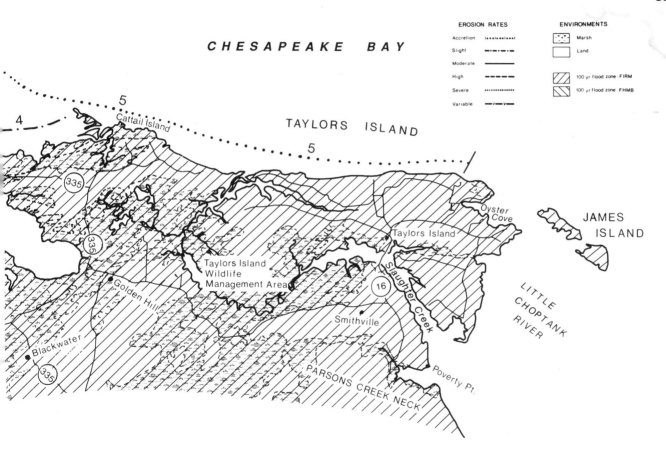

Table 4.7 Erosion rates: Dorchester County, Maryland (Taylors and Hooper islands)

Segment number	Erosion rate
1	Slight to moderate; some accretion in sheltered areas
2	Slight to moderate
3	Slight to moderate; some severe
4	Accretion to slight
5	High to severe

Table 4.8 Erosion rates: Dorchester County, Maryland (Little Choptank and Choptank rivers)

Segment number	Erosion rate
1	Severe
2	Slight to moderate
3	Moderate to severe
4	Accretion to slight
5	Moderate to high
6	Accretion to slight
7	High to severe
8	Accretion to slight
9	High to severe
10	Accretion to slight (except where open to wave attack from bay; then moderate)
11	High to severe
12	Variable; accretion to high
13	High to severe
14	Accretion to slight
15	Variable; slight to high
16	Variable (accretion to slight in back of bay and protected areas; moderate where exposed)
17	Slight
18	Moderate to high
19	Slight; stable where structured

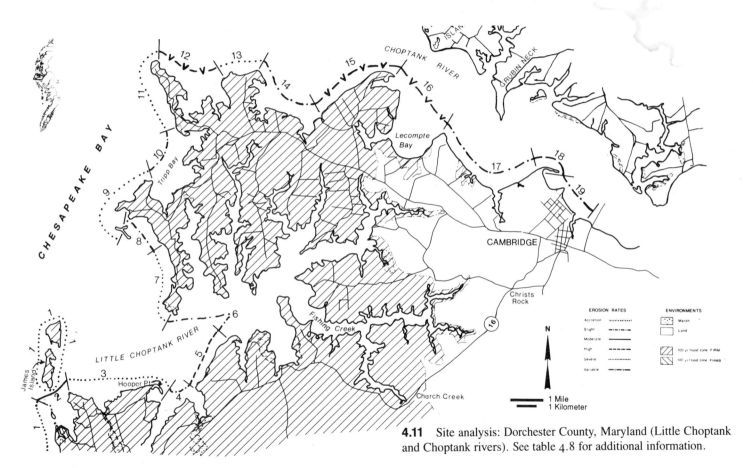

4.11 Site analysis: Dorchester County, Maryland (Little Choptank and Choptank rivers). See table 4.8 for additional information.

102 Selecting a site

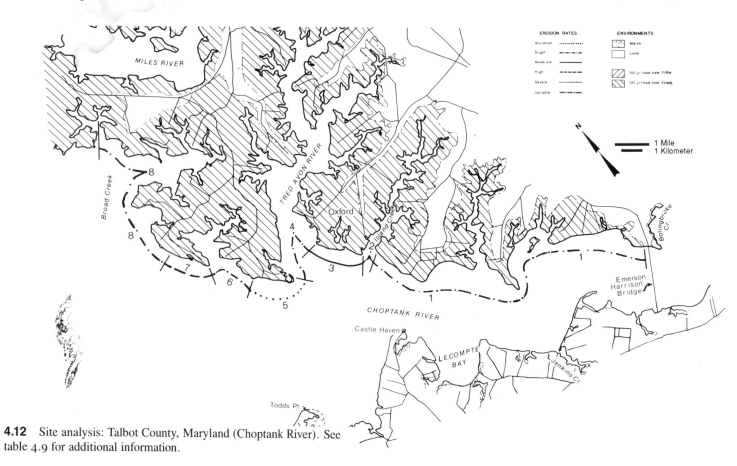

4.12 Site analysis: Talbot County, Maryland (Choptank River). See table 4.9 for additional information.

Table 4.9 Erosion rates: Talbot County, Maryland (Choptank River)

Segment number	Erosion rate
1	Accretion to slight
2	Slight
3	Slight to moderate
4	Accretion to slight
5	High to severe
6	Accretion to slight
7	Moderate to high
8	Accretion to slight

Lower Western Shore, Virginia: Northumberland, Lancaster, Middlesex, Mathews, Gloucester, and York counties, Hampton Roads, and Norfolk

The erosion rates along the southern shore of the lower Potomac River and adjacent Chesapeake Bay shore have historically been high. This is partially due to the exposure to northeast storm waves. Some of the most severe erosion has occurred on both sides of the mouth of Cod Creek, where retreat has been 10 feet per year to the west, and 6 feet per year to the east. Several homes are presently threatened by erosion along this outer shore, in spite of "shore-defense" structures. The shorelines along the numerous creeks are most often lined with fringing salt marsh, and tend to have low erosion rates. Many of the land areas adjacent to the creek shores are at higher elevations, which offer a degree of flood protection.

Between the Potomac and Rappahannock Rivers (fig. 4.31, table 4.27; fig. 4.32, table 4.28; and fig. 4.33, table 4.29), portions of Northumberland and Lancaster counties border on the Chesapeake Bay. This region is undergoing rapid development varying from agriculture to waterfront recreation. The shore areas tend to be low and are dissected by creeks, thus making them highly susceptible to flooding. As in most parts of the Bay, the small creeks are lined with fringing marshes and tend to be stable. The Bay-front portions of this area are exposed to high wave energy, but in general, shoreline retreat is not a severe problem. Average shore retreat is less than 2 feet per year. The more sheltered shoreline along the

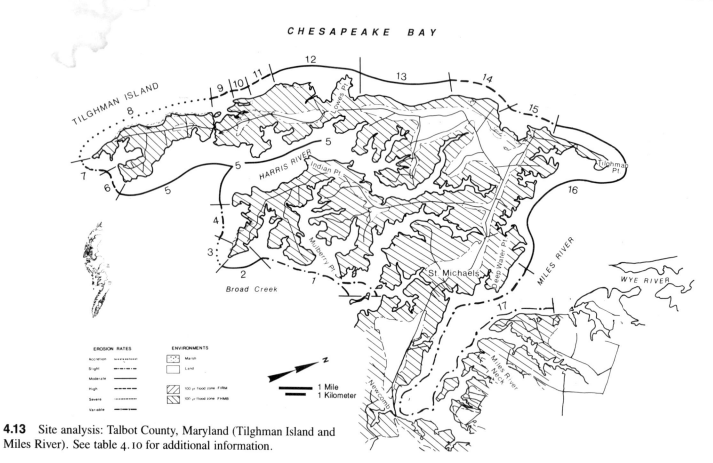

4.13 Site analysis: Talbot County, Maryland (Tilghman Island and Miles River). See table 4.10 for additional information.

Table 4.10 Erosion rates: Talbot County, Maryland (Tilghman Island and Miles River)

Segment number	Erosion rate
1	Accretion to slight
2	Slight to moderate
3	High to severe
4	Accretion to slight; some moderate rates near Nelson Point and Change Point
5	Slight to moderate; some minor accretion
6	Moderate to high
7	Accretion to slight
8	Moderate to severe
9	Moderate to high
10	High
11	Moderate to high
12	Slight to moderate; some high
13	Slight to moderate
14	Slight to high
15	Accretion to slight
16	Slight to moderate
17	Accretion to slight

4.14 Cambridge, Maryland.

106 Selecting a site

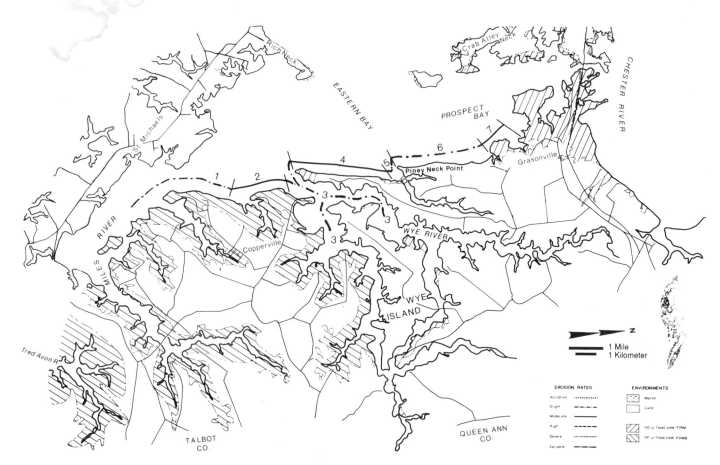

Rappahannock averages only 0.6 feet per year of erosion, though individual extremes include the area north of Windmill Point (8 feet per year) and south of Indian Creek (6 feet per year).

From Gwynn Island to the south shore of the Rappahannock River are portions of Gloucester, Mathews, and Middlesex counties (fig. 4.34, table 4.30; fig. 4.35, table 4.31). Erosion rates are highly variable along the shore and tend to be related to the degree of exposure to open stretches of water, and hence to storm waves. High rates of retreat occur at Stingray Point at the confluence of two rivers, where the shore has receded an average of over 6 feet per year. Shore retreat in areas protected from storm waves is often related to storm surges, which occasionally elevate water levels sufficiently to erode the bases of bluffs. After the storm recedes, slumping of the bluffs may take place, thus causing shore retreat.

The portion of Mathews County facing the open Bay (fig. 4.35) is experiencing severe rates of shore retreat (table 4.31). North of New Point Comfort, a long barrier beach is eroding at rates that are typically over 8 feet per year. Similar to open-ocean barrier islands, these low-lying areas are periodically overwashed by storms. Sand is deposited on the marsh behind the islands while the beach retreats landward. As a result, rollover or migration of the barrier is taking place. Both the width of the island and the dune elevations are quite small in this region because the supply of

4.15 Site analysis: Talbot and Queen Annes counties, Maryland (Eastern and Prospect bays). See table 4.11 for additional information.

Table 4.11 Erosion rates: Talbot and Queen Annes counties, Maryland (Eastern and Prospect bays)

Segment number	Erosion rate
1	Accretion to slight
2	Slight to moderate
3	Accretion to slight
4	Slight to moderate
5	Slight (estimated)
6	Primarily slight; some moderate rates near Piney Neck Point
7	Slight to moderate

sand is low. The flood potential is also high, so property owners and residents should be on the alert for storms. This is an area that requires prudent development.

Gwynn Island, just north of the barrier islands, is also eroding at rates up to 7 feet per year. Most of the shore is now protected by riprap, bulkheads, and groins.

The Mobjack Bay shoreline consists of a large embayment that opens into the Chesapeake (fig. 4.36, table 4.32). As there is a broad expanse of water in front of Mobjack Bay, wave energies can be quite high. Areas of highest erosion are found in the vicinity of Ware Neck Point and Robins Neck with average shore retreat

4.16 Site analysis: Kent Island, Maryland. See table 4.12 for additional information.

Table 4.12 Erosion rates: Kent Island, Maryland

Segment number	Erosion rate
1	Primarily slight
2	Accretion to slight
3	Slight to moderate
4	Slight to moderate; lowest rates in protected areas like Crab Alley Creek; some high rates at headlands
5	Moderate to high
6	Accretion to slight
7	Variable (slight to high); some headlands like Butts Neck have severe
8	Mostly slight to moderate; some areas of high
9	Mostly slight to moderate; some accretion
10	Variable (slight to high); some severe erosion near Kent Point
11	Slight to moderate
12	Moderate to severe
13	Variable; severe toward northern end (Love Point)
14	Moderate to high
15	Slight
16	Slight to moderate
17	Accretion to slight
18	Moderate to high
19	Accretion to slight

110 Selecting a site

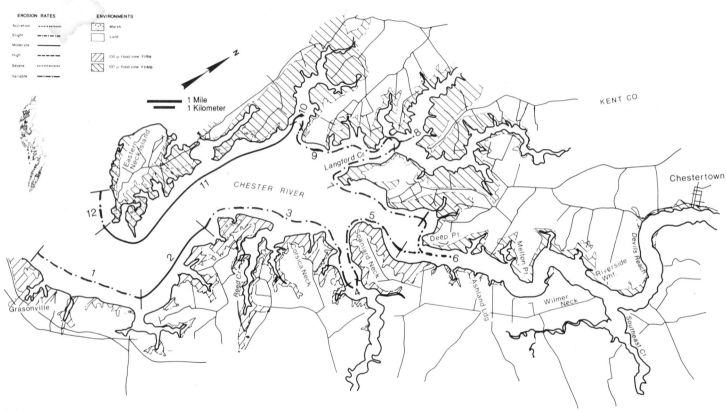

4.17 Site analysis: Chester River, Maryland. See table 4.13 for additional information.

Table 4.13 Erosion rates: Chester River, Maryland

Segment number	Erosion rate
1	Predominantly slight
2	Predominantly slight to moderate
3	Accretion to slight
4	Accretion to slight (estimated)
5	Accretion to slight
6	Accretion to slight (estimated)
7	Accretion to slight
8	Accretion to slight (estimated)
9	Accretion to slight
10	Accretion to slight (estimated)
11	Slight to moderate
12	Slight

exceeding 2 feet per year. As in most areas of the Chesapeake Bay, stable portions of the shore tend to be in creeks and the upper portions of the rivers. Most of the shoreline in Mobjack Bay has at least a narrow fringing marsh, which tends to increase the stability of the shoreline. Property owners should preserve the marsh in order to maintain stability.

The outer shore of York County is largely an undeveloped, low-lying area undergoing erosion (fig. 4.37, table 4.33). York Point, at the northern entrance to Poquoson River, is a notable exception to this undisturbed coastal region. Here, the salt marsh has been filled and narrow canals excavated to provide waterfront lots for private homes. Although this practice has been common in the lagoons associated with the barrier islands facing the open Atlantic, this is a rare example in the Chesapeake Bay. At the present time, salt marshes are carefully protected along all of the Bay shoreline and destruction of marsh by development is rarely tolerated.

North of Hampton Roads the north-south trending shoreline from Northend Point to Old Point Comfort, is, for all practical purposes, a barrier island beach. The southern two-thirds of this shoreline is occupied by Buckroe Beach and Fort Monroe, which is armored with bulkheads and groins. This barrier beach has been retreating at the rate of 4 to 6 feet per year. Fort Monroe, incidentally, is the only active military facility in the United States surrounded by a moat.

The Hampton Roads region is a fully developed shoreline (fig. 4.38, table 4.34). Virtually all of the shore has been stabi-

112 Selecting a site

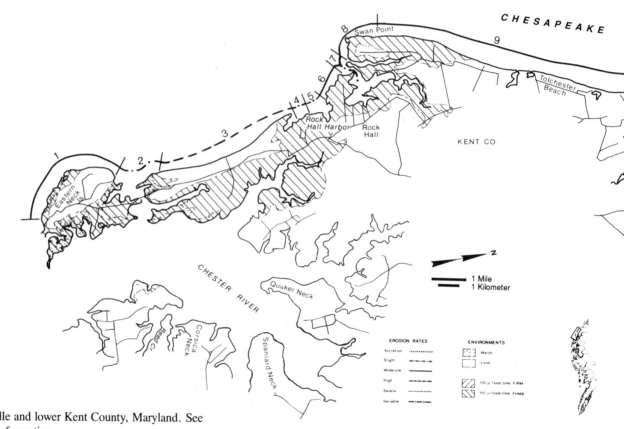

4.18 Site analysis: middle and lower Kent County, Maryland. See table 4.14 for additional information.

Selecting a site 113

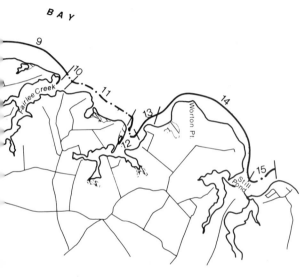

Table 4.14 Erosion rates: lower and middle Kent County, Maryland

Segment number	Erosion rate
1	Slight to moderate
2	Accretion to slight
3	Moderate to high
4	Slight
5	Slight to accretion
6	Slight to moderate
7	Slight
8	Moderate
9	Slight to moderate; some small areas of accretion
10	Slight (estimated)
11	Accretion to slight
12	Slight (estimated)
13	Moderate
14	Slight to moderate
15	Slight

114　Selecting a site

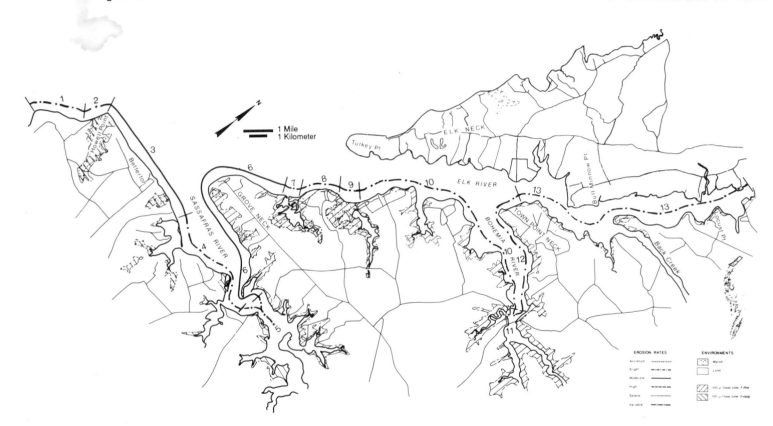

4.19　Site analysis: upper Kent and Cecil counties, Maryland. See table 4.15 for additional information.

Table 4.15 Erosion rates: upper Kent and Cecil counties, Maryland

Segment number	Erosion rate
1	Accretion to slight
2	Slight
3	Slight to moderate
4	Accretion to slight
5	Accretion to slight (estimated)
6	Slight to moderate
7	Accretion to slight
8	Stable to slight
9	Slight
10	Accretion to slight
11	Accretion to slight (estimated)
12	Accretion to slight
13	Accretion to slight (estimated)

lized by a variety of bulkheads and other structures. There are almost no natural beaches and, because of the stabilization, almost no shore retreat. Fringe marsh still flourishes in a few protected creek areas.

The southern shore of the Chesapeake Bay (fig. 4.39, table 4.35) is directly exposed to the high wave energy that enters Chesapeake Bay from the open ocean or is generated in the lower Bay. Consequently, the beaches in the lower Bay closely resemble those along the Atlantic. They are often wide and sandy and have high rates of longshore drift of sand.

Cape Henry is a spit formed from the northward-flowing sand along the ocean shore of Virginia Beach that is trapped at the mouth of the Bay. The sand on these beaches is transported by waves and currents passing into the Bay. This beach system ends to the west at Willoughby Spit. In the early part of this century, Willoughby Spit retreated to the south about 200 feet and became 500 feet longer. Beginning in the 1920s, groins were constructed to halt the erosion. The spit now has a series of over thirty groins, and long-term shore retreat averages one-quarter foot per year.

The open-ocean shore of Virginia Beach, Sand Bridge, and Virginia's barrier islands is treated separately in chapter 5.

116 Selecting a site

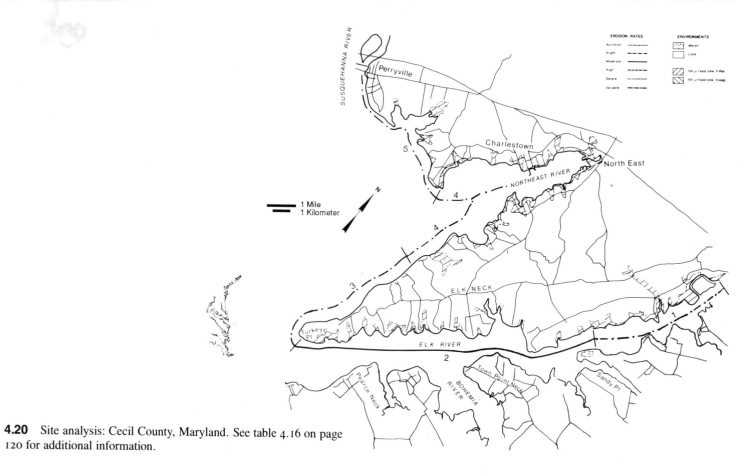

4.20 Site analysis: Cecil County, Maryland. See table 4.16 on page 120 for additional information.

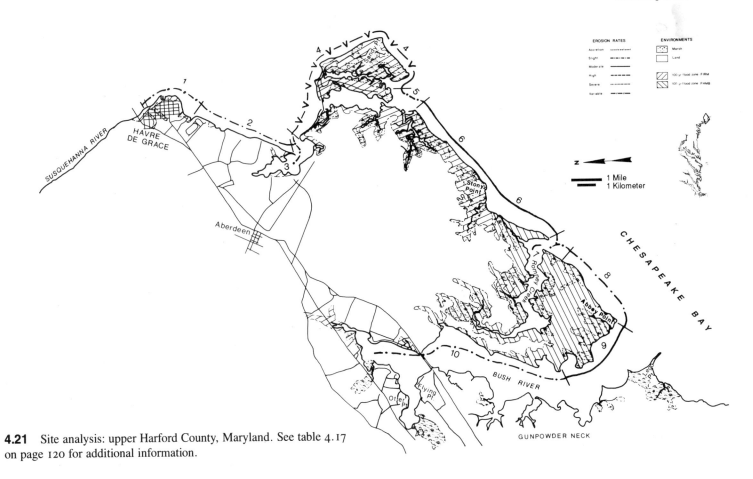

4.21 Site analysis: upper Harford County, Maryland. See table 4.17 on page 120 for additional information.

4.22 Site analysis: lower Harford and upper Baltimore counties, Maryland. See table 4.18 on page 120 for additional information.

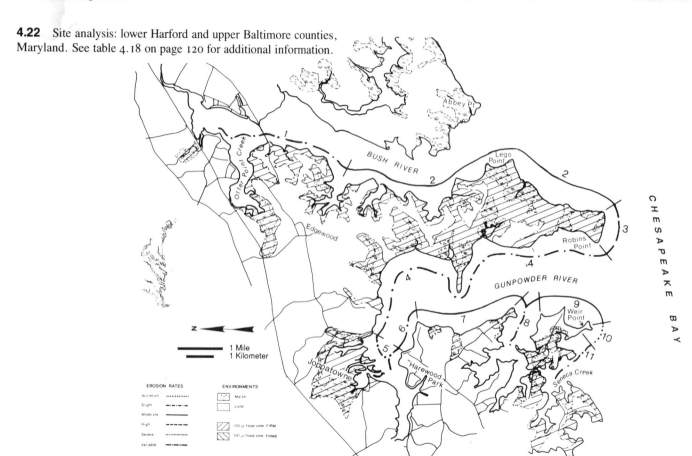

Selecting a site

4.23 Site analysis: Baltimore County, Maryland. See table 4.19 on page 120 for additional information.

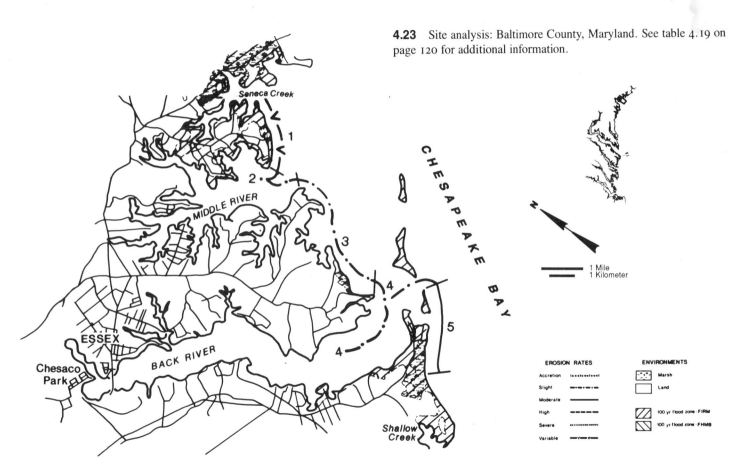

Table 4.16 Erosion rates: Cecil County, Maryland

Segment number	Erosion rate
1	Accretion to slight (estimated)
2	Slight to moderate
3	Accretion to slight
4	Slight
5	Accretion to slight

Table 4.17 Erosion rates: upper Harford County, Maryland

Segment number	Erosion rate
1	Predominantly slight
2	Accretion to slight
3	Accretion to slight (estimated)
4	Variable; slight in protected areas; moderate to high where shore faces open bay
5	Accretion to slight (estimated)
6	Primarily slight to moderate; high near Stony Point
7	Accretion to slight (estimated)
8	Slight (estimated)
9	Slight to moderate
10	Slight (estimated)

Table 4.18 Erosion rates: lower Harford and upper Baltimore counties, Maryland

Segment number	Erosion rate
1	Accretion to slight
2	Slight to moderate
3	Moderate to high
4	Primarily slight
5	Slight (estimated)
6	Stable to slight
7	Slight to moderate
8	Primarily accretion to slight
9	Primarily slight to moderate
10	Severe
11	Slight

Table 4.19 Erosion rates: Baltimore County, Maryland

Segment number	Erosion rate
1	Variable; accretion to slight in protected areas; slight to moderate along reaches exposed to the open bay
2	Accretion to slight
3	Primarily accretion to slight
4	Accretion to slight
5	Slight to moderate

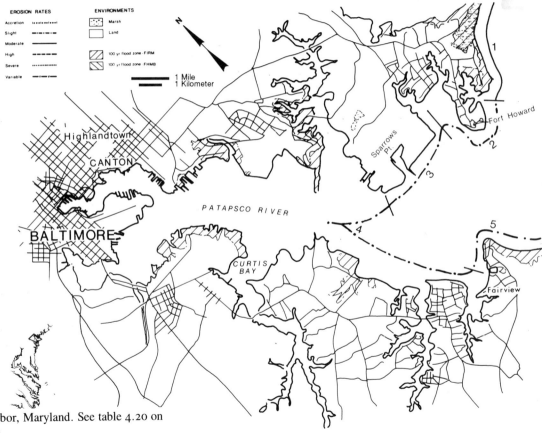

4.24 Site analysis: Baltimore Harbor, Maryland. See table 4.20 on page 124 for additional information.

122 Selecting a site

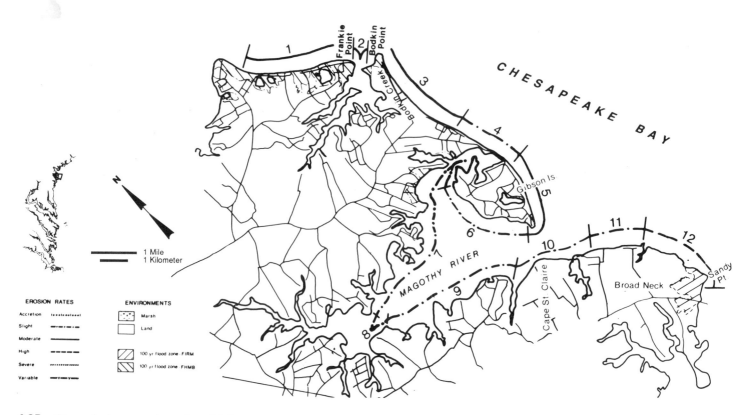

4.25 Site analysis: upper Anne Arundel County, Maryland. See table 4.21 on page 124 for additional information.

4.26 Site analysis: middle Anne Arundel County, Maryland. See table 4.22 on page 124 for additional information.

Table 4.20 Erosion rates: Baltimore Harbor, Maryland

Segment number	Erosion rate
1	Slight to moderate
2	Accretion to slight
3	Accretion due to infilling
4	Accretion where filled in; some slight
5	Accretion to slight

Table 4.21 Erosion rates: upper Anne Arundel County, Maryland

Segment number	Erosion rate
1	Primarily slight to moderate; some high rates near Frankie Point
2	Accretion to slight (estimated)
3	Mostly slight to moderate; some high near Bodkin Point
4	Slight
5	Primarily slight to moderate
6	Slight
7	Accretion to slight
8	Accretion to slight (estimated)
9	Primarily slight; some accretion
10	Primarily slight; some moderate; accretion to slight in embayments
11	Slight
12	High

Table 4.22 Erosion rates: middle Anne Arundel County, Maryland

Segment number	Erosion rate
1	Moderate to high
2	Slight to moderate
3	Primarily high to severe; slight in protected embayments
4	Slight
5	Accretion to slight (estimated)
6	Accretion to slight
7	Slight to moderate
8	Accretion to slight (estimated)
9	Primarily slight to moderate
10	High to severe
11	Accretion to slight
12	Primarily accretion to slight; some areas have been filled in
13	Primarily accretion to slight
14	Moderate to high
15	Slight to moderate
16	Accretion to slight
17	Primarily slight to moderate
18	Accretion to slight
19	Moderate to high
20	Accretion to slight

Selecting a site 125

4.27 Site analysis: lower Anne Arundel County, Maryland. See table 4.23 for additional information.

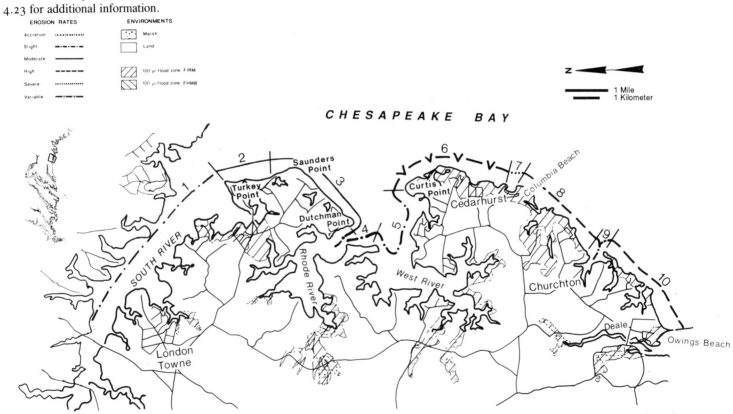

126 Selecting a site

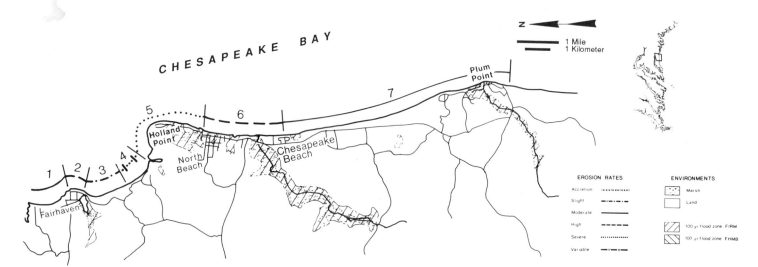

4.28 Site analysis: lower Anne Arundel and upper Calvert counties, Maryland. See table 4.24 for additional information.

Table 4.23 Erosion rates: lower Anne Arundel County, Maryland

Segment number	Erosion rate
1	Accretion to slight
2	Slight to moderate
3	Primarily slight to moderate; high near Sanders Point
4	Slight (estimated)
5	Predominantly accretion to slight
6	Variable; accretion to slight in areas not exposed to wave attack; high to severe where exposed to main stem of Chesapeake Bay
7	High to severe
8	Predominantly moderate to high; some severe
9	Accretion to slight
10	Slight to high

Table 4.24 Erosion rates: lower Anne Arundel and upper Calvert counties, Maryland

Segment number	Erosion rate
1	Slight to moderate
2	High
3	Accretion to slight
4	Accretion to stable
5	High to severe
6	Primarily moderate to high
7	Primarily slight to moderate; some accretion just north of Plum Point

Table 4.25 Erosion rates: middle and lower Calvert County, Maryland

Segment number	Erosion rate
1	Slight to moderate
2	Mostly slight to moderate; high in some isolated areas
3	Moderate
4	Accretion to slight
5	Slight to moderate
6	Northern segment has high rate of erosion; southern portion shows accretion
7	Primarily slight
8	North of Cove Point has high to severe; south of Cove Point shows long-term accretion
9	Primarily slight to moderate; some high
10	Slight
11	Slight to moderate; some accretion at Drum Point
12	Slight
13	Accretion to slight; some areas have been filled

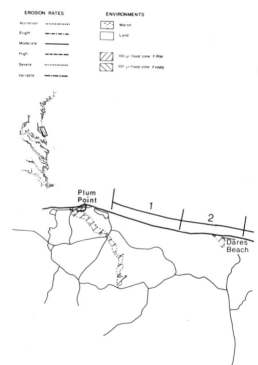

4.29 Site analysis: middle and lower Calvert County, Maryland. See table 4.25 for additional information.

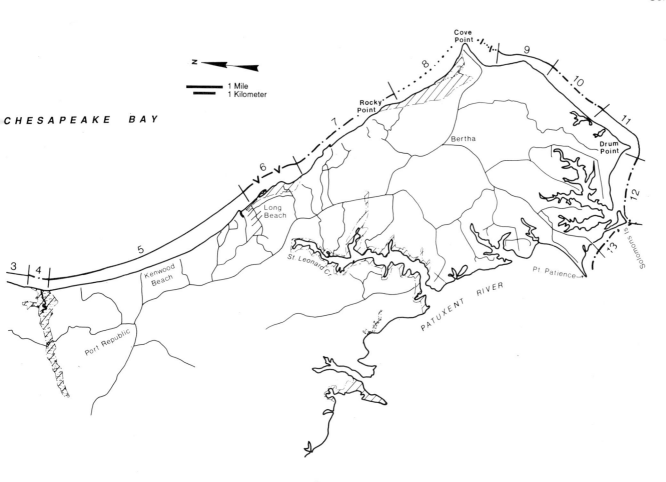

4.30 Site analysis: Saint Marys County, Maryland. See table 4.26 for additional information.

Table 4.26 Erosion rates: Saint Marys County, Maryland

Segment number	Erosion rate
1	Accretion to slight
2	Variable; high to severe by Cedar Point; slight to the south
3	Slight to moderate
4	High to severe
5	Primarily slight to moderate; some areas of accretion at Pt. No Point
6	Accretion to slight (estimated)
7	Variable; high to severe
8	Accretion
9	Predominantly high to severe
10	Slight; some accretion at Point Lookout
11	Moderate to high
12	Primarily slight to moderate
13	Moderate to severe
14	Accretion to slight (estimated)
15	Moderate to high
16	Accretion to slight
17	Slight to moderate
18	Accretion to slight (estimated)
19	Accretion to slight
20	Primarily accretion to slight
21	Moderate to high
22	Accretion to slight (estimated)
23	Accretion to slight
24	Moderate to high

132 Selecting a site

4.31 Site analysis: upper Northumberland County, Virginia. See table 4.27 for additional information.

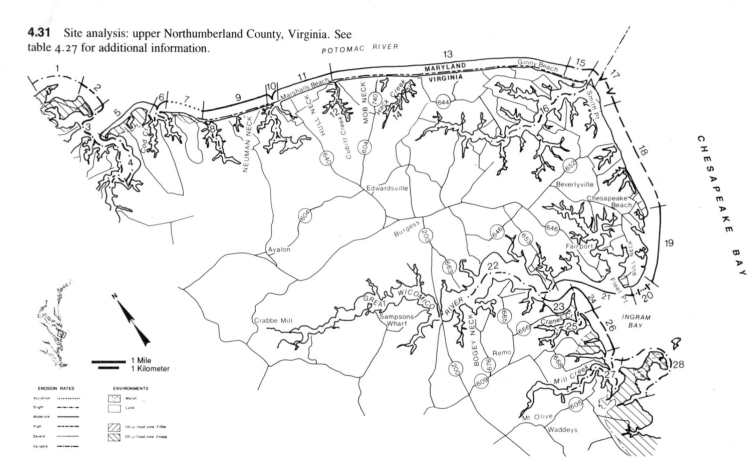

Table 4.27 Shoreline types and erosion rates: upper Northumberland County, Virginia

Segment number	Shoreline type	Erosion rate	Segment number	Shoreline type	Erosion rate
1	Marsh margin	Slight	15	Beach	Slight
2	Beach (man-modified)	High	16	Fringe marsh	Slight
3	Fringe marsh with bluff	Slight	17	Beach with dune	Variable
4	Beach with low bluff	Slight	18	Beach with low bluff (man-modified)	High
5	Beach with low bluff	Moderate	19	Beach (man-modified)	Moderate
6	Fringe marsh	Slight	20	Spit with dune	Accretion
7	Beach (man-modified)	Severe	21	Fringe marsh (man-modified)	Moderate
8	Fringe marsh	Slight	22	Fringe marsh	Slight
9	Beach with low bluff	Moderate	23	Beach with low bluff	Moderate
10	Fringe marsh	Slight	24	Beach (man-modified)	Moderate
11	Beach (man-modified)	Moderate	25	Fringe marsh	Slight
12	Fringe marsh	Slight	26	Beach	High
13	Beach with low bluff (man-modified)	Moderate	27	Fringe marsh	Slight
14	Fringe marsh	Slight	28	Marsh margin and fringe marsh	Slight

4.32 Site analysis: lower Northumberland and Lancaster counties, Virginia. See table 4.28 for additional information.

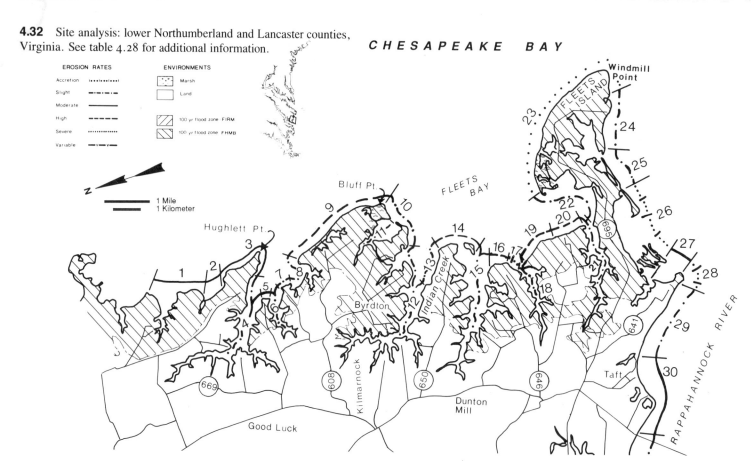

Table 4.28 Shoreline types and erosion rates: lower Northumberland and Lancaster counties, Virginia

Segment number	Shoreline type	Erosion rate	Segment number	Shoreline type	Erosion rate
1	Fringe marsh	Moderate	16	Beach with dune (man-modified)	High
2	Beach	Moderate	17	Beach (man-modified)	Accretion
3	Marsh margin and sandy barrier	Moderate	18	Fringe marsh (man-modified)	Slight
4	Fringe marsh	Slight	19	Beach (man-modified)	High
5	Beach	Moderate	20	Beach (man-modified)	High
6	Fringe marsh (man-modified)	Slight	21	Fringe marsh	Slight
7	Beach (man-modified)	Slight	22	Beach	Slight
8	Fringe marsh	Slight	23	Beach fringe marsh (man-modified)	Slight
9	Beach	High	24	Man-modified	Slight
10	Marsh margin, sandy barrier	Slight	25	Beach with dune	Slight
11	Marsh margin, fringe marsh	Slight	26	Beach (man-modified)	Accretion
12	Fringe marsh	Slight	27	Man-modified	Moderate
13	Beach (man-modified)	Slight	28	Man-modified	Accretion
14	Beach	High	29	Beach with high bluff	Slight
15	Fringe marsh beach (man-modified)	Slight	30	Beach with high bluff	Moderate

4.33 Site analysis: lower Lancaster County, Virginia. See table 4.29 for additional information.

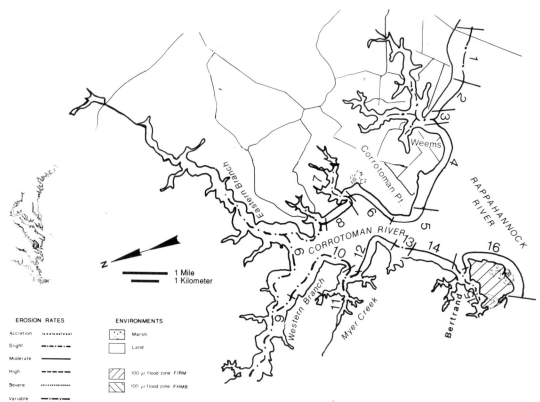

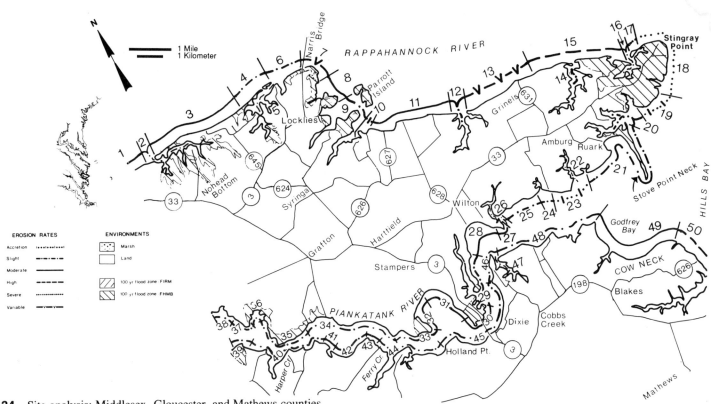

4.34 Site analysis: Middlesex, Gloucester, and Mathews counties, Virginia (Rappahannock to the Piankatank River). See table 4.30 for additional information.

Table 4.29 Shoreline types and erosion rates: lower Lancaster County, Virginia

Segment number	Shoreline type	Erosion rate
1	Beach with high bluff (man-modified)	Slight
2	Beach with low bluff	Moderate
3	Fringe marsh (man-modified)	Slight
4	Fringe marsh (man-modified)	Moderate
5	Beach	Moderate
6	Fringe marsh	Moderate
7	Fringe marsh	Slight
8	Man-modified	Moderate
9	Fringe marsh	Slight
10	Beach with low bluff	Slight
11	Fringe marsh	Slight
12	Beach (man-modified)	Moderate
13	Fringe marsh	Slight
14	Beach with low bluff	Moderate
15	Fringe marsh	Slight
16	Beach marsh (man-modified)	Moderate

Table 4.30 Shoreline types and erosion rates: Middlesex, Gloucester, and Mathews counties, Virginia (Rappahannock to the Piankatank River)

Segment number	Shoreline type	Erosion rate
1	Beach (man-modified)	Moderate
2	Fringe marsh	Slight
3	Beach	Moderate
4	Beach	Slight
5	Fringe marsh	Slight
6	Beach, fringe marsh	Slight
7	Beach (man-modified)	Variable
8	Beach, fringe marsh	Moderate
9	Fringe marsh	Slight
10	Beach (man-modified)	Accretion
11	Beach (man-modified)	Moderate
12	Fringe marsh	Slight
13	Beach with high bluff (man-modified)	Variable
14	Fringe marsh with high bluff	Slight
15	Beach (man-modified)	High
16	Fringe marsh (man-modified)	Slight
17	Beach (man-modified)	Slight
18	Beach (man-modified)	Severe

Segment number	Shoreline type	Erosion rate
19	Beach (man-modified)	Accretion
20	Fringe marsh (altered)	Slight
21	Beach (man-modified)	Slight
22	Fringe marsh	Slight
23	Beach with high bluff	Accretion
24	Fringe marsh, high bluff (man-modified)	Slight
25	Beach with low bluff (man-modified)	Slight
26	Fringe marsh	Slight
27	Beach (man-modified)	Moderate
28	Beach with high bluff	Moderate
29	Beach, fringe marsh (man-modified)	Slight
30	Fringe marsh (man-modified)	Moderate
31	Beach with high bluff (man-modified)	Moderate
32	Fringe marsh	Moderate
33	Fringe marsh	Slight
34	Fringe marsh beach	Slight
35	Beach (man-modified)	Slight
36	Fringe marsh	Slight
37	Beach (man-modified)	Slight
38	Fringe marsh	Slight
39	Fringe marsh (man-modified)	Slight
40	Fringe marsh	Slight
41	Fringe marsh	Moderate
42	Beach with high bluff	Slight
43	Beach	Moderate
44	Fringe marsh	Slight
45	Beach with high bluff	Slight
46	Beach	Slight
47	Fringe marsh	Slight
48	Beach with high bluff	Slight
49	Beach with high bluff	Medium
50	Beach with high bluff	High

140 Selecting a site

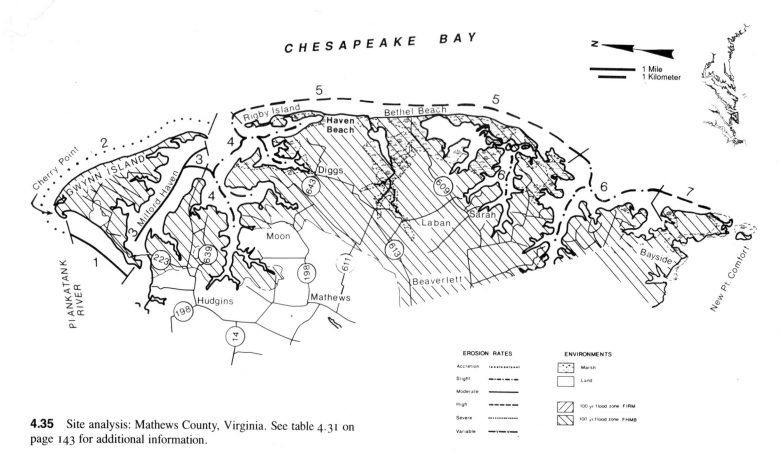

4.35 Site analysis: Mathews County, Virginia. See table 4.31 on page 143 for additional information.

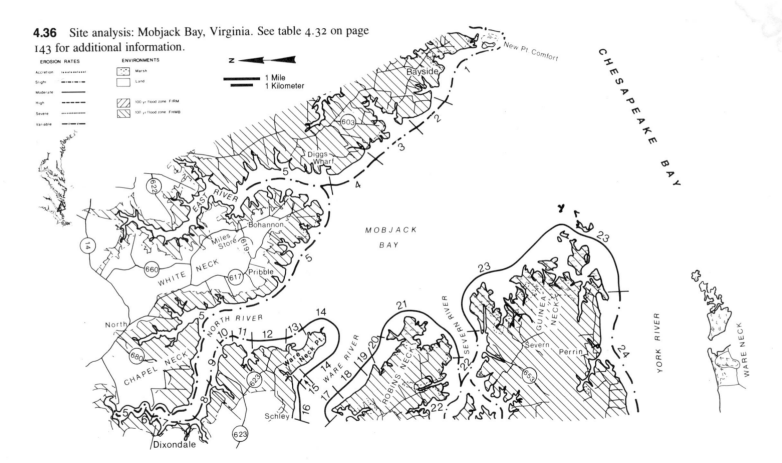

4.36 Site analysis: Mobjack Bay, Virginia. See table 4.32 on page 143 for additional information.

142 Selecting a site

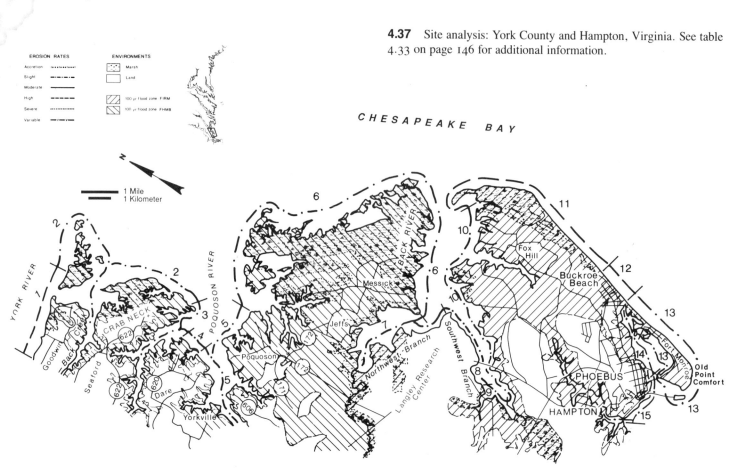

4.37 Site analysis: York County and Hampton, Virginia. See table 4.33 on page 146 for additional information.

Table 4.31 Shoreline types and erosion rates: Mathews County, Virginia

Segment number	Shoreline type	Erosion rate
1	Man-modified	Moderate
2	Beach (man-modified)	Severe
3	Beach, fringe marsh (man-modified)	Moderate
4	Fringe marsh	Slight
5	Beach dune	High
6	Fringe marsh	Slight
7	Beach	High

Segment number	Shoreline type	Erosion rate
8	Fringe marsh	Moderate
9	Fringe marsh	Slight
10	Beach	Slight
11	Fringe marsh	Slight
12	Fringe marsh (man-modified)	Moderate
13	Beach, fringe marsh	Slight
14	Fringe marsh	Moderate
15	Fringe marsh	High
16	Fringe marsh (man-modified)	Moderate
17	Beach with dune	Moderate
18	Fringe marsh (man-modified)	Moderate
19	Beach	Moderate
20	Fringe marsh	Moderate
21	Beach, marsh	Moderate
22	Fringe marsh	Slight
23	Beach, fringe marsh	Moderate
24	Beach (man-modified)	Slight

Table 4.32 Shoreline types and erosion rates: Mobjack Bay, Virginia

Segment number	Shoreline type	Erosion rate
1	Fringe marsh beach	Slight
2	Fringe marsh	High
3	Fringe marsh	Slight
4	Beach	High
5	Fringe marsh	Slight
6	Fringe marsh	Moderate
7	Fringe marsh	Slight

144 Selecting a site

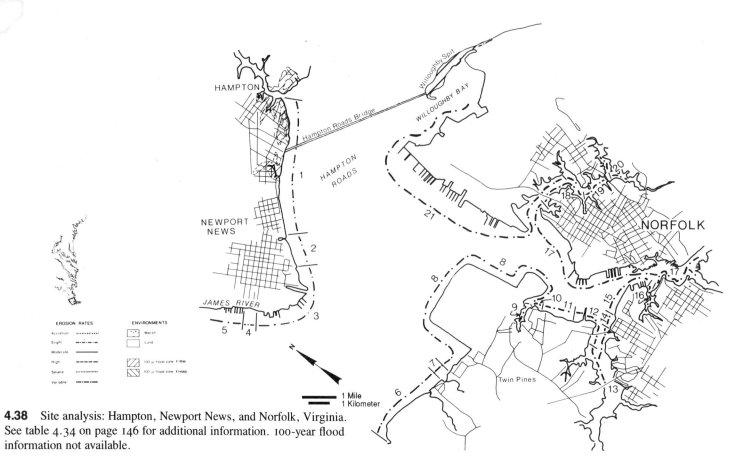

4.38 Site analysis: Hampton, Newport News, and Norfolk, Virginia. See table 4.34 on page 146 for additional information. 100-year flood information not available.

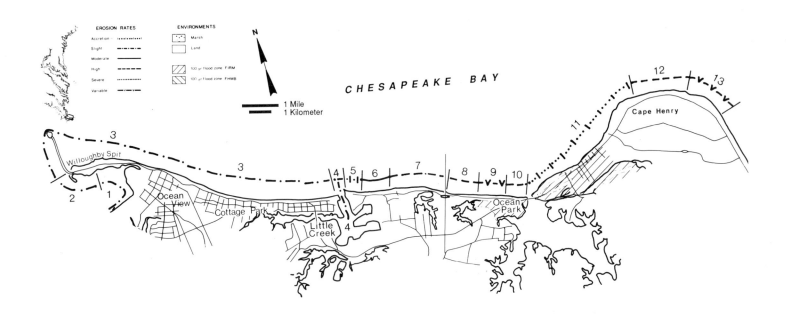

4.39 Site analysis: Willoughby Spit to Cape Henry, Virginia. See table 4.35 on page 147 for additional information.

Table 4.33 Shoreline types and erosion rates: York County and Hampton, Virginia

Segment number	Shoreline type	Erosion rate
1	Beach (man-modified)	High
2	Fringe marsh, marsh margin	Slight
3	Fringe Marsh	Slight
4	Beach	Slight
5	Fringe Marsh	Slight
6	Beach dune, fringe marsh	Slight
7	Fringe marsh	Slight
8	Man-modified	Slight
9	Fringe marsh	Slight
10	Fringe marsh (man-modified)	Slight
11	Beach dune	High
12	Man-modified	High
13	Man-modified	Slight
14	Fringe marsh	Slight
15	Man-modified	Slight

Table 4.34 Shoreline types and erosion rates: Hampton, Newport News, and Norfolk, Virginia

Segment number	Shoreline type	Erosion rate
1	Man-modified	Slight
2	Beach (man-modified)	Slight
3	Man-modified	Slight
4	Beach with low bluff	Slight
5	Man-modified	Slight
6	Man-modified	Slight
7	Fringe marsh (man-modified)	Slight
8	Man-modified	Slight
9	Fringe marsh	Slight
10	Man-modified	Slight
11	Beach	Slight
12	Man-modified	Slight
13	Fringe marsh (man-modified)	Slight
14	Beach (man-modified)	Slight
15	Man-modified	Slight
16	Fringe marsh (man-modified)	Slight
17	Man-modified	Slight
18	Fringe marsh (man-modified)	Slight

Selecting a site 147

Table 4.35 Shoreline types and erosion rates: Willoughby Spit to Cape Henry, Virginia

Segment number	Shoreline type	Erosion rate
1	Beach (man-modified)	Slight
2	Man-modified	Slight
3	Beach (man-modified)	Slight
4	Man-modified fringe marsh	Slight
5	Beach dune	Accretion
6	Beach dune	Moderate
7	Beach dune	High
8	Beach (man-modified)	High
9	Beach (man-modified)	Variable
10	Beach	High
11	Beach dune	Accretion
12	Beach dune (man-modified)	High
13	Beach dune	Variable
19	Man-modified	Slight
20	Fringe marsh (man-modified)	Slight
21	Man-modified	Slight

5 The open-ocean shoreline

Delmarva's barrier islands

Virginians can take pride in the fact that their open-ocean barrier-island shoreline (fig. 5.1) is among the most beautiful and the least developed in the nation. Almost the entire stretch of islands on the Delmarva Peninsula south of Ocean City, Maryland, is in federal, state, or Nature Conservancy hands. These islands stand in stark and beautiful contrast to crowded Fenwick Island. A thriving tourist industry is based on the Virginia islands and in the future more and more people will have the privilege of visiting what is becoming a scarce American commodity: a pristine barrier island.

It is a good thing that most of these islands are not developing. Most are experiencing a very rapid rate of shoreline retreat—unusually fast even by the standard of Atlantic barrier islands elsewhere. If development had proceeded here as on other East or Gulf Coast barriers, a great deal of environmental and economic damage would have taken place by now. Imagine, if you will, a whole series of communities lining the shore of the outer Delmarva Peninsula with even more severe erosion problems than Ocean City, Maryland, or Sandbridge, Virginia.

The erosion experienced by these islands is exemplified by the experiences of Broadwater, a town located on the six-mile-long Hog Island, earlier this century. In the early 1930s over 300 people lived on the island in 50 homes, with one hotel, one church, and four stores. The island already had a long history of erosion. Three thousand feet of beach retreat occurred at the southern tip of the island between 1871 and 1910. Then the great August 1933 storm struck. Although property damage was only moderate, 16 people died in the storm. After the storm most people moved off the island, some taking their homes with them. By 1942 only 13 buildings remained, many others having been claimed by the relentless erosion. In 1943 the post office closed, and the town of Broadwater on Hog Island ceased to exist. In later years local watermen reported finding tombstones from the Broadwater cemetery in their nets. According to the *Ledger-Star* newspaper in Norfolk, at high tide Hog Island is now but a thin strip of sand.

Cobb Island (fig. 5.1), first settled by Nathan Cobb, a New England shipbuilder, experienced a similar history. In the late 1800s the island had only 50 year-round residents, but a booming tourist industry. It boasted a luxury hotel, and among its distinguished visitors was Jefferson Davis! In 1868 accommodations cost a whopping $3 per day or $18 per week. Unfortunately, a series of storms destroyed everything in the 1890s. By 1897 Cobb Island was essentially abandoned.

The open-ocean shoreline   149

The 1890s were a turbulent time for other coastal communities as well. A bustling community on Shackleford Banks (Diamond City), near Cape Lookout in North Carolina was also abandoned in the late 1890s, although subsequent erosion has not taken away the homesites as is the case at Cobb Island.

More recently, NASA's Wallops Island missile facility has been the site of massive efforts to stop shoreline erosion. Seawalls, bulkheads, and groins, now crumbling under the forces of the sea, have failed to arrest the movement of the shoreline. At one point artificial seaweed was used in front of Wallops Island.

The problems experienced by the early resorts on Virginia barrier islands result from the tendency of such islands to change shape, size, and orientation in response to changing natural conditions (such as supply of sand, shape of the offshore sea floor, and wave energy). In fact, barrier islands everywhere are responding to the rise in sea level by going through changes that are part of a long-range process known as island migration (see discussion in chapter 2). Barrier islands typically erode on the ocean side while material is deposited on the back side by storm overwash or other processes. This builds the island in a landward direction. The result is an island rolling over on itself like a tank tread. What was once the back side becomes the ocean side. Mud from bay-side salt marshes is commonly found on Virginia's barrier islands (including Sandbridge to the south). This marsh mud outcropping on ocean beaches is proof of island migration.

5.1 Index map of the Delmarva barrier islands and Virginia Beach.

Perhaps the most spectacular example in America of island migration can be seen on northern Assateague Island in Maryland. In 1933 and 1934, the U.S. Army Corps of Engineers built a set of jetties at the south end of Ocean City. The jetties cut off the supply of sand to Assateague and the island immediately began to migrate landward. Now the northern end of Assateague Island, which was once in line with the beachfront of Ocean City, has nearly completely moved off its original position. In the future it may migrate right over the Intracoastal Waterway in the bay behind the island.

Our recent understanding of how barrier islands migrate has made a great deal of difference in the management policies of national seashores. In the old days (1930s) it was assumed that barrier-island erosion meant that the islands would soon disappear. Thus, the frantic, successful dune-building projects on North Carolina's Outer Banks, just south of the Virginia line, were assumed to be essential to the survival of the islands. Now the National Park Service has decided to let the islands roll on. Dune building with snow fences is no longer attempted.

By a combination of luck, disastrous experiences of the early settlers, local economics, and the foresight of the Nature Conservancy, most of the Virginia barrier islands have been preserved for future generations without being a tax burden on the present generation. These islands hold the promise of being a great state treasure for our great-grandchildren.

The exception to this may occur at Cedar Island. Currently, there are approximately 18 homes on Cedar Island that are widely spread out. Despite the extremely high erosion rates (greater than 15 feet per year in places), there has been a proposed development of 50 to 100 more lots. At the time of this writing, the issue has not been resolved.

South of the mouth of the Chesapeake Bay, a different story unfolds. Virginia Beach is one of the most extensively developed Atlantic barrier shorelines in the United States (fig. 5.2). It is at one and the same time a major economic boost to the state economy and a major economic drain because of the cost of holding the shoreline in place against all that nature throws at it.

Virginia Beach: facing the sea

The open-ocean shoreline of Virginia Beach is a different kettle of fish, as the old saying goes. Much of this book is concerned with the Chesapeake Bay shoreline, which differs greatly from the open ocean in many ways. First and foremost, the energy of the waves striking the shoreline in the Bay is usually much less than that on the open ocean. Normal wave conditions at Virginia Beach are more like storm-wave conditions in the Bay. Storm waves on Virginia Beach are more severe than anything ever seen in the Bay.

There are other important differences too. For one thing, the sand supply to the open-ocean beaches is much larger. For another, the beaches of Virginia Beach are magnets that draw 1.5 million people a year who spend over $300 million (1986). This provides the city with $10 million in taxes. Recreational beaches in the Bay (excluding Norfolk, which is revitalizing its beaches at

Ocean View and Willoughby) are generally small in comparison and are used mainly by small numbers of people. In other words, preservation of beaches in the Bay, although an important environmental concern, is not a major economic factor in the life of Bay residents.

Virginia Beach residents face the same problem that millions of people are faced with on the entire Atlantic barrier-island shoreline —from Long Island's south shore to Miami Beach, Florida. The most overriding and certainly the most visible problem is stopping shoreline erosion to preserve rapidly narrowing beaches. How to save threatened buildings without destroying the all-important beaches is a topic that is constantly on the mind of local residents and local officials.

In addition, Virginia Beach residents who live near the open-ocean beach are faced with the problem of direct onslaught of hurricane winds and waves. There is the problem of overwashing storm waves. Communities like Sandbridge face an additional hazard of flooding during storms because water is pushed into the bay behind the island. Strong winds with accompanying waves quickly sort out the poorly built from the well-built houses. Unfortunately, even well-built buildings may be severely damaged from debris flying from houses of less-careful neighbors. Evacuation before a big storm is essential, and if it is not done well ahead of time, residents may find their way blocked either by rising storm waters or by the cars of thousands of people trying to exit at the same moment. Many North Carolina vacationers were trapped on barrier islands by the swift arrival of Hurricane Charley (August 1986).

5.2 Virginia Beach's shoreline is one of the most heavily developed U.S. barrier shorelines. (A) Virginia Beach area near Ruddee Inlet. (B) Virginia Beach near the Chesapeake Bay Bridge and Tunnel.

Fortunately, Charley left little damage in his wake; next time we may not be so lucky.

Living on or visiting Virginia's Atlantic coast can be a pleasant experience, but prudent individuals will understand the environment they live in and will plan ahead for the inevitable problems caused by the clash of man and the sea on a dynamic shoreline.

Virginia Beach has a long and stormy history. Particularly memorable storms are the August 1933 hurricane and the great Ash Wednesday storm, a nor'easter that struck in March of 1962. These big storms cut highways, destroyed seawalls, and damaged buildings. Smaller storms have destroyed walls since 1962, and the damage from the next big one may be greater than ever. On Virginia Beach, as on so many other American shorelines, there is now much more to damage.

A less spectacular change in Virginia Beach has been the gradual narrowing and near-loss of the beach in front of the city. In 1952 the state created the Virginia Beach Erosion Commission, which has watchdogged the erosion problem ever since. The city has determined that its beaches are essential for the local economy and so a great deal of money has been spent "saving" them. One of the first projects the commission became involved in was a beach-nourishment project. Over a million cubic yards of sand was placed on beaches that were in bad shape. According to the *Ledger-Star* newspaper (June 1, 1986) the commission proudly reported its work to then Governor John S. Battle: "An unparalleled story of man conquering the sea to preserve a nationally famous beach resort playground." To maintain this victory, Virginia Beach has continued to dump sand on the beach.

Because the city also values its buildings, it has allowed extensive seawall construction and other efforts to reduce erosion. For instance, in 1973 the city installed an experimental offshore breakwater called Potter's Fence. The breakwater, installed a few hundred feet offshore, between 17th and 20th Streets, consisted of a series of steel A-frames. In 1973 the city agreed to pay $108,000 for the device when it appeared to be working a few months after installation. It is questioned today whether the device ever actually worked. In 1984 the city agreed to pay an additional $60,000 to have the fence removed.

Currently, Virginia Beach, which has the oldest continuous beach-nourishment program on the East Coast, spends about $1.5 million every year to transfer sand from Lynnhaven Inlet, Rudee Inlet, and private sand pits to the beaches. This cost, which is shared by federal, state, and local taxpayers, can be assumed to be a minimum annual cost for future years. Roughly 250,000 cubic yards of sand (sometimes more, sometimes less) is placed on a 3.3-mile stretch of beach each year from Rudee Inlet north to 49th Street by the Erosion Commission. Some of this sand is pumped in; some of it (typically about 150,000 cubic yards) is trucked in (fig. 5.3). Since each dump truck carries 10 cubic yards it is easy to see why these vehicles are an important part of the local beach scene. The dominant natural direction of transport along the beach here is south to north. This sand is undoubtedly helping to widen the beach of northern Virginia Beach.

At the time of this writing, a major beach-restoration plan that

5.3 Dumping of sand on Virginia Beach from trucks helps maintain this resort beach.

was formulated by the U.S. Army Corps of Engineers is being considered. In order to preserve the beach and create hurricane protection, a concrete seawall from Rudee Inlet to 57th Street needs to be constructed, the 100-foot-wide beach maintained, and sand dunes between 57th and 89th Street enlarged (25-foot crest). This proposal would cost around $36 million.

Sandbridge

Sandbridge (fig. 5.4) is located on a barrier island. At this particular instant in geologic time, it is actually part of a long peninsula extending down to Oregon Inlet in North Carolina. However, historically, numerous inlets have existed along this shoreline and more may cut through in future storms.

Sandbridge is a community with severe erosion problems. Because erosion rates are high and construction has been undertaken too close to the beach, a number of buildings are threatened. Some buildings have already been condemned because their septic-tank systems are exposed. At this writing, about a dozen bulkheads are in place in front of some homes.

Although the policy of the City of Virginia Beach is not to allow the building of new bulkheads, a two-block area was exempted by a bill signed in 1985 by former Governor Robb. Residents owning property in this critically eroding beachfront area had been denied a permit by the local Wetlands Board to build bulkheads to protect their property from the sea. The Wetlands Board cited strong evidence by coastal scientists that indicates bulkheads accelerate the erosion of the adjacent beach when waves are reflected off the bulkheads carrying sand farther offshore than normal. Both the Virginia Marine Resources Commission and the Virginia Beach Circuit Court had upheld the Wetlands Board decision before Robb signed the bill. The incident has led to heated debate over local versus state control of the Sandbridge area, and the rights of individual property owners versus the protection of the beach.

Privately financed bulkheads in front of individual homes here have had varied results. The Army Corps recently investigated the possibility of replenishing the beach at Sandbridge, but concluded that it was not feasible from a public-benefit standpoint. This decision is being contested by local private-property owners.

On the bay side of Sandbridge, finger canals have been cut into the low-lying marsh to furnish waterfront homesites with access to North Bay. Here, a number of homeowners are being threatened by erosion. It is incorrect to think that this area is exempt from erosion problems.

Coastal development

The Virginia Beach shoreline stretches from the entrance to Chesapeake Bay to the North Carolina border. Within this stretch of shoreline is much federal- and state-owned property that will not be developed commercially. Virginia Beach's Chesapeake Bay shoreline was reviewed in chapter 4 (fig. 4.39). Along the Atlantic-facing shore there are two reaches of shoreline where development has proceeded and where individuals may still buy or build homes. These are Virginia Beach proper and Sandbridge. Between the two is a U.S. Navy reservation, and south of Sandbridge is a National Wildlife Refuge and a state park. The Virginia Beach developed shoreline is roughly seven miles long; Sandbridge is situated on about five miles of shoreline between the naval reservation and the wildlife refuge.

The stretch of shoreline from Croatan Beach, south of Rudee

5.4 Sandbridge, Virginia.

Inlet, to the southern boundary of Fort Story is a moderately hazardous area in which to live in terms of shore erosion and flooding problems. The aspects of this stretch of shoreline that are hazardous to inhabitants include: flooding potential in storms due to low elevations; exposure of shorefront buildings to direct wave activity, which is enhanced by the narrowness of the beach; and possible difficulties evacuating the area due to large numbers of people and low elevations of some portions of escape roads and bridge approaches.

Sandbridge is considered to be a high-risk area in terms of erosion and flooding due to: high flood potential from both the bay side and the open-ocean side; extremely high erosion rates; danger from direct wave attack due to low elevations; and potential evacuation difficulty due to the low elevation of escape roads. However, the small number of people living here will reduce possible traffic tie-ups during storm evacuation.

6 Coastal land use and the law

During the colonial period there were few legal restraints on shoreline development. The Chesapeake Bay, as a natural resource, was considered to be free for the use of property owners in any reasonable way that they saw fit. "*Riparian rights*," an English common-law interpretation, generally assured these owners the right to take reasonable measures to protect their land against erosion and gain access to navigable water, as well as other uses normally associated with shoreline ownership. These rights, while having undergone considerable interpretation over time, still form the basis for modern law in the Bay area.

It was well after independence from colonial rule that the legislatures in Maryland and Virginia took steps to codify shoreline rights. Until then, all changes depended on court interpretations of the law. Some of the most important legislation affecting shoreline property has been passed in recent years. Especially significant from the viewpoint of property owners and developers are measures taken in both Maryland and Virginia to preserve the wetlands bordering much of the Bay's shoreline. The conventional wisdom that marshlands are a "waste" of land has been replaced by an understanding that the vitality of the entire Bay is influenced by the retention of wetlands in their natural state. Valuable fisheries in both Maryland and Virginia could be seriously damaged if the wetlands were destroyed. Similarly, the relationships between prudent land use and the integrity of the Bay system are being recognized.

Although Maryland and Virginia seem to have similar historical roots and are subject to the same federal regulation, important differences exist between the laws of the two states. These differences may in some cases be traced to the original colonial charters, and in other cases to different economic and political circumstances. For example, there are marked differences between states in permitting shoreline development, even though there is substantial agreement on methodology and goals. Both states have legal commitments to wetlands protection, but they go about implementing them differently because of their different legal environments. The same is true for regulating land use, shoreline construction, and water resources, and for mitigating coastal hazards.

Anyone purchasing and developing land at the shore or taking actions that affect the coastal zone should be aware of the regulations, codes, and ordinances that govern his actions. The following introduction to federal, state, and local programs is a starting point for the coastal property owner. However, the proper officials should be consulted for the most up-to-date and complete information on coastal laws and regulations. Appendix B provides addresses of some of the agencies that can provide additional information, necessary permit applications, and advice. Appendix C provides references that will give greater detail on specific regulations.

Federal programs

National Flood Insurance Program

The National Flood Insurance Act of 1968 as amended by the Flood Disaster Protection Act of 1973 was passed to encourage prudent land-use planning and to minimize property damage in flood-prone areas, including the coastal zone. Local communities must adopt ordinances to reduce flood risks in order to qualify for the National Flood Insurance Program (NFIP). The NFIP provides an opportunity for property owners to purchase flood insurance that generally is not available from private insurance companies.

In the past, the NFIP has been subsidized and has grown to become a large federal liability. As of August 31, 1981, more than 1.918 million flood insurance policies valued at $97,972 billion had been sold nationwide. Coastal counties had 1.165 million of these policies valued at $64,667 billion.

The Federal Emergency Management Agency (FEMA) Disaster Assistance Program Division serves as an advisory agency for the reduction of impacts due to natural hazards (e.g., flooding, landslides, earthquakes), and exerts some regulatory control to reduce future property damage. Under the authority of the federal Disaster Relief Act of 1974 (Public Law 93-288), the agency evaluates potential hazards and determines plans to mitigate the effects of such hazards. Reduction of loss due to flooding is specifically addressed under the authority of the Inter-Agency Agreement for Nonstructural Flood Damage Reduction, and Executive Order 11-988: Flood Plain Management, which designates FEMA as the lead agency to determine actions that will reduce the impact of flooding.

The initiative for qualifying for NFIP rests with the community, which must contact FEMA. Any community may join NFIP (unless excluded by the Coastal Barrier Resources Act), provided that it requires permits for all proposed construction and other development within the flood zone and ensures that construction materials and techniques are used that minimize potential flood damage. Initially, a community enters the "emergency phase" of the NFIP. The community is provided with a Flood Hazard Boundary Map (FHBM) which serves as a preliminary delineation of flood-hazard areas. During this phase, the federal government makes a limited amount of flood insurance available, charging subsidized premium rates for all existing structures and/or their contents, regardless of the flood risk.

To enter the "regular program" of the NFIP, the community must adopt and enforce flood-plain management ordinances that at least meet the minimum requirements for flood-hazard reduction as set by FEMA. The advantage of entering the regular program is that more insurance coverage is available. All new structures are rated on an actual risk (actuarial) basis, which may mean higher insurance rates in potential high-hazard areas, but generally results in a savings for development within "*A-zones*" (areas flooded in a 100-year coastal flood, but less subject to turbulent wave action).

In the Chesapeake Bay region, FEMA has produced detailed Flood Insurance Rate Maps (FIRMs) for most of the coastal area. The FIRM shows flood elevations and flood-hazard zones, including velocity zones (*V-zones*) for coastal areas where wave action is an additional hazard during flooding. The FIRM identifies base flood elevations (BFE), establishes special flood-hazard zones, and provides a basis for managing flood plains and establishing insurance rates.

FEMA maps commonly use the "100-year flood" as the base flood elevation to establish regulatory requirements. Persons unfamiliar with hydrological data sometimes mistakenly take the "100-year flood" to mean a flood that occurs once every 100 years. In fact, a flood of this magnitude could occur in successive years, or even twice in the same year. A 100-year flood is a level of flooding that has a 1 percent statistical probability of occurring in a given year. Having flood insurance makes good sense.

Flood elevations in the V-zones of coastal counties take into account the additional hazard of storm waves atop still-water flood levels. In V-zones, elevation requirements are adjusted, usually 3 to 6 feet above still-water flood levels, to minimize wave damage. The insurance rates are also higher. When your insurance agent submits an application for a building within a V-zone, he must attach an elevation certificate verifying the postconstruction elevation of the first floor of the building.

Existing FEMA regulations stipulate protection of dunes and vegetation in the V-zones, but implementation of this requirement by local communities has not always been strong. The existing requirements of the NFIP do not address other hazards of migrating shorelines, e.g., shoreline erosion and shifting of inlets.

The insurance-rate structure provides incentives of lower rates if buildings are elevated above minimum federal requirements. Eligibility requirements generally vary among pole houses, mobile homes, and condominiums. Flood-insurance coverage is provided for structural damage as well as for contents.

Most coastal communities are now covered under the regular program. To determine if your community is in the NFIP and for additional information, contact your local property agent, insurance agent, NFIP's servicing contractor, or the NFIP Region III Office (appendix B).

Most lending institutions and community-planning, zoning, and building departments are aware of the flood-insurance regulations and can provide assistance. It would be wise to confirm such information with appropriate insurance representatives. Any authorized insurance agent can write and submit a NFIP policy application. For more information request a copy of "Questions and Answers on the National Flood Insurance Program" from FEMA. The addresses of their offices are given in appendix B.

Before buying a building or structure in the coastal zone an individual should ask certain basic questions:

— Is the community I'm locating in covered by the emergency phase or regular program of the NFIP?

— Is the building site above the 100-year–flood level? Is the site located in a V-zone? V-zones are higher-hazard areas and can pose serious problems.
— What are the minimum elevation and structural requirements for my building?
— What are the limits of my coverage?

Nature has a way of exceeding human expectations. Storm waves coming from the right direction at just the right time on a high spring tide may cause flood levels that exceed the predicted levels of the flood maps. Homeowners should regard the requirements necessary to obtain flood insurance as minimal, and go beyond those requirements when elevating and flood proofing their structures.

Some flood-insurance facts

1. Flood insurance offers the potential flood victim a less expensive and broader form of protection than would be available through a postdisaster loan.
2. Flood insurance is a separate policy from homeowner's insurance. Know the difference between a homeowner's policy and a flood-insurance policy. From the standpoint of water damage, homeowner's insurance often covers only structural damage from wind-driven rain.
3. Flood insurance typically covers losses that result from the general and temporary flooding of normally dry land, the overflow of inland or tidal waters, and the unusual and rapid accumulation of runoff or surface water from any source.

Check to see if your location has been identified as flood prone on the FIRM. If you are located in a flood-prone area you must purchase flood insurance in order to be eligible for most forms of federal or federally related financial, building, or acquisition assistance—that is, VA and FHA mortgages, Small Business Administration loans, and similar assistance programs. To locate your property on the FIRM map, see your insurance agent. Also keep in mind: (1) You need a separate policy for each structure. (2) If you own the building, you can insure the structure and contents, or contents only, or structure only. (3) If you rent the building, you need only insure the contents. A separate policy is required to insure the property of each tenant.

For flood-insurance purposes a condominium unit that is a traditional town house or row house is considered as a single-family dwelling and may be separately insured.

Mobile homes are eligible for coverage if they are on foundations, whether or not permanent, and regardless of whether the wheels are removed either at the time of purchase or while on the foundation.

Structures and other items that are not eligible for flood insurance are: (1) travel trailers and campers; (2) fences, retaining walls, seawalls, septic tanks, and outdoor swimming pools; (3) gas and liquid storage tanks, wharves, piers, bulkheads, growing crops, shrubbery, land, livestock, roads, or motor vehicles.

One insurance broker cannot charge you more than another for the same flood-insurance policy because the rates are subsidized and set by the federal government.

There is a five-day waiting period from the date of application until the coverage becomes effective.

Hurricane evacuation

The Disaster Relief Act of 1974 authorized FEMA to establish disaster-preparedness plans in cooperation with local communities and states. Hurricane evacuation is a critical problem on barrier islands and coastal flood plains. Due to heavy concentrations of population in areas of low topography, narrow roads, and vulnerable bridges and causeways, plus limited hurricane-warning capability (possibly 12 hours or less), it may be difficult to evacuate completely before a hurricane strikes. Persons living in the lower Bay or surrounding Atlantic coastal region should contact municipal, county, or state agencies for hurricane evacuation maps and information (see appendixes A and B).

Coastal Barrier Resources Act of 1982

Recognizing the serious hazards, costs, and problems with federally subsidized development of barrier islands, the U.S. Congress passed the Coastal Barrier Resources Act (Public Law 97-348) in October 1982. The purpose of this federal law is to minimize loss of human life and property; wasteful expenditure of federal taxes; and damage to fish, wildlife, and other natural resources from incompatible development along the Atlantic and Gulf coasts. The act covers 186 designated areas, covering 700 miles of undeveloped barrier beaches in the United States.

Specifically, the act prohibits the expenditure of federal funds, including loans and grants, for the construction of infrastructures that encourage barrier-island development in the 186 designated areas. These infrastructures include roads, bridges, water-supply systems, waste-water-treatment systems, and erosion-control projects. Any new structure built on these barrier islands after October 1, 1983, is not eligible for federal flood insurance. Certain activities and expenditures are permissible under the act. The act does not prohibit private development on the designated barrier islands, but it passes the risks and costs of development from taxpayers to owners.

The Coastal Barrier Resources Act includes the Virginia open-ocean coastal units of Assawoman Island, Cedar Island, Little Cobb Island, and Fishermans Island. No Maryland coastal areas were included under the original act. In 1985 the Department of the Interior proposed to extend the Coastal Barrier Resources System to include other parts of the Delmarva barrier islands and parts of Chesapeake Bay. At the time of this writing, the proposed additions to the Coastal Barrier Resources System had not become effective.

Navigable waterways

Federal jurisdiction in the tidewater area is carried through the U.S. Army Corps of Engineers. The corps' authority to regulate projects in "navigable waters of the United States" dates back to the Federal River and Harbor Act of 1899, which was principally aimed at maintaining the navigability of those waters. The corps was given additional responsibilities under section 404 of the Federal Water Pollution Control Act of 1972. Its authority was later extended to all "waters of the United States," including wetlands (which may, arguably, be termed nonnavigable), and was further defined by the Clean Water Act of 1977. Most projects affecting navigable waterways and tidal wetlands, including dredging, fill work, and construction, are subject to the corps' approval. Many types of projects may be exempt or covered by a general permit. It is wise to check with the local corps district office (listed in appendix B) to learn of specific requirements before planning work in a wetland or waterway.

Federal Coastal Zone Management Act

In 1972 the federal government passed the Coastal Zone Management Act (CZMA) to provide assistance in establishing state coastal-zone–management programs. The act was advisory in nature, providing financial incentives by making federal funding available to assist in the development and implementation of state programs. It outlines a general framework for state coastal management whereby existing or new legislation can qualify a state for participation.

In 1977 the state of Maryland submitted a coastal zone management plan (CZMP), which was approved in 1978. This CZMP makes extensive use of existing state laws and legal authorities through a "networking" concept. The Tidewater Administration (Coastal Resources Division) of the Maryland Department of Natural Resources is the lead organization charged with implementing the provisions of the CZMP. The plan encompasses defined activities within the coastal zone. These activities include beach-erosion control, wetlands and critical-areas management, activities in navigable waters, and development such as construction impacting coastal resources. Details concerning the Maryland CZMP can be obtained by contacting the Tidewater Administration, Maryland Department of Natural Resources, Tawes State Office Building, Annapolis, MD 21401.

The commonwealth of Virginia recently approved a CZMP. However, important coastal-resources laws have been in effect for years in Virginia, including the Coastal Primary Sand Dune Protection Act and the Wetlands Act.

Local and state laws

Local zoning laws in Maryland and Virginia, as in other states, are of principal importance in defining allowable land use. State laws concerning issues such as development rely for enforcement on local zoning offices. Therefore, local and state laws must comply.

Because local planning offices are charged in part with the enforcement of state regulations, they are excellent places to ask for information concerning proposed development in a given area. Additional information can be obtained from state or federal government offices dealing with subjects like wetlands protection, erosion control, flood protection, etc.

Some specific activities for which restrictions may apply or permits may be required are discussed below. These activities are controlled at the municipal or county level unless otherwise stated, and jurisdiction can be determined by inquiring at local offices (e.g., planning office, building inspector, health department).

Critical Area Law (Maryland)

The Chesapeake Bay Critical Area Law, passed in 1984 by the state of Maryland, regulates land use and development around Maryland's portion of the Bay. "Critical area" refers to all land and water areas of Chesapeake Bay and its tributaries within 1,000 feet of the landward boundary of state or private wetlands and heads of tide. Previous restrictions on development or regulatory laws were largely devoted to wetlands areas. With the implementation of the Critical Area Law, all the land in Maryland immediately surrounding the Bay is regulated.

The purpose of this new legislation is to minimize adverse impact to water quality and natural habitats of Chesapeake Bay. Results of a major study of the Bay by the U.S. Environmental Protection Agency indicated that the cumulative effects of human activity have caused a deterioration of the health of the Bay. In order to decrease the levels of pollutants, nutrients, and toxins entering the Bay system, the state government has taken a number of regulatory steps, including the Critical Area Law. This law severely limits new development of much of the undeveloped shoreline around the Bay. In addition, its restrictions include the requirement of buffer strips around existing agricultural fields, around livestock feeding or watering areas, and on all undeveloped shorelines. Also, limits have been set on commercial forest harvesting within the critical area.

The Chesapeake Bay Critical Area Commission, which is directed to develop the standards and criteria of the Critical Area Law, has defined three categories of land use in Chesapeake Bay. These are (1) Intensely Developed Areas (IDAS), (2) Limited Development Areas (LDAS), and (3) Resource Conservation Areas (RCAS).

IDAS are areas where residential, commercial, institutional, or industrial development predominates and little natural habitat remains over at least 20 continuous acres (or the entire upland portion of a municipality within the Critical Area—whichever is less). Also, in order to be considered intensely developed, a stretch of coast has to have at least one of the following criteria: housing density equals or exceeds four dwellings per acre; industrial, institutional, or commercial uses of land are concentrated; or public sewer and water collection and distribution systems are currently serving the area (and more than three dwellings per acre exist).

LDAS include areas where only low or moderate development has occurred, natural habitats exist, and there has been no substantial alteration or degradation of run-off. In addition, LDAs have one of the following characteristics: dwellings from one unit per five acres up to four per acre; the area is not dominated by agriculture, wetland, forest, barren land, surface water, or open space; the area has the characteristics of IDAs but fewer than 20 acres; or the area has public water and/or sewer lines.

RCAs are characterized by nature-dominated or resource utilization activities (e.g., wetlands, forests, abandoned fields, agriculture, fisheries, or aquaculture). In addition, housing density is less than one dwelling per five acres, or the dominant land use is agriculture, wetland, forest, barren land, surface water, or open space.

Although all the land-use types around the Bay have restrictions, most development restrictions are for RCAs, so new development is directed toward IDAs. Some moderate development may occur in LDAs. New development within RCAs is restricted to one dwelling per 20 acres. There are exceptions to these regulations, which will allow development in an area equal to 5 percent of a county's RCAs (excluding tidal wetlands or federally owned areas); but only 50 percent of this development can be in the RCAs. Also there are a number of "grandfather" clauses and methods for "intrafamily transfers." Excluding these exceptions, the Critical Area Law encourages future intense development outside the critical area.

Although the Critical Area Law was developed at the state level, primary responsibility for developing and implementing critical-area resource-protection programs is charged to local governments.

Because of the profound effect the Critical Area Law has on development of shorelines around Maryland, potential and existing homeowners in critical areas should become thoroughly familiar with the new regulations.

Questions concerning developing land in specific areas around Chesapeake Bay should be directed to local government officials and planning offices. Information can also be obtained directly from the Chesapeake Bay Critical Area Commission, Tawes State Office Building, D-4, Annapolis, MD 21401. Of specific interest are the reports "Subtitle 15 Chesapeake Bay Critical Area Commission Criteria for Local Critical Area Program Development: Final Regulations, Synopsis of Criteria Development" and "A Guide to the Chesapeake Bay Critical Area Criteria" (reference 125, appendix C).

Pending land-use regulation (Virginia)

At the time of this writing Virginia had a number of laws in review that would have a similar effect on land use and development around Virginia's Chesapeake Bay and Atlantic Coast. As in Maryland, the best place to start inquiring prior to developing or altering a shoreline area is local planning or government offices.

Zoning and subdivision (all areas)

County and municipal offices maintain zoning maps describing the types of development determined to be desirable for and allowable in a given area. Zoning variances can sometimes be obtained when conflicts arise, but they may not be feasible in all cases. This option cannot be fairly treated here, and should best be discussed with an authority familiar with the local laws. Remember that an area zoned for a particular use may be undevelopable for reasons other than zoning restrictions.

Subdivision restrictions may limit the scope of development allowed within a particular property. These limits should be determined in advance as well.

Many of the shoreline areas of the Chesapeake have seen rapid development in recent years. In some cases zoning and subdivision restrictions may reflect a local desire to minimize the impact of a too-rapid growth rate on old, established communities.

Building permits (all areas)

Any structure located in a community that is part of the National Flood Insurance Program will be required to meet certain structural standards for flood proofing set by the federal government. These represent minimum standards that are required in order that the community may qualify for NFIP.

Additional requirements may exist at the local level. It is best to check with municipal and county planning offices, which will be able to give advice on the applicability of NFIP and other requirements.

Sedimentation control (all areas)

Construction and development can increase sediment loads in storm-water runoff, which can be harmful to sensitive wetlands and Bay waters. Depending on the nature of the development project, it may be necessary to obtain a sediment-control permit before beginning construction.

Water supply and sewerage (all areas)

Building permits are normally withheld until adequate provisions for water and sewerage needs are met. Sometimes, plans for water and sewerage must be approved before subdivision and sale of lots for private development are allowed.

Although subject to state regulation, these requirements are administered locally through state health officers and county and municipal authorities.

General wetlands protection

The tidal wetlands of Chesapeake Bay are an important yet vulnerable part of the estuarine environment. These wetlands have a level of diversity and abundance of life not found in many other environments. Their importance as a wildlife habitat and breeding area makes them essential to the ecology of the entire Bay.

Development of waterside real estate has already destroyed thousands of acres of wetlands bordering the Bay and, if unchecked, will continue to do so. As the population of the Bay area continues to increase, more of the shoreline areas are likely to become attractive to development. The need to ensure that this development is of minimal detriment to the Bay is viewed as a high priority in both Maryland and Virginia.

Regulatory measures aimed at protecting wetlands ecosystems have been enacted at federal and state levels in the Bay area. The Critical Area Law addresses this issue in Maryland. Many of the regulations concerning wetlands ecosystems are of very recent origin and may seem contrary in spirit to traditional riparian rights. The basic structure of riparian rights is still intact, though it is true that recent legislation has resulted in perhaps the biggest modification of these rights in over a century.

The states of Maryland and Virginia have passed wetlands laws that require permits in addition to those required by federal law. In fact, approval of permits by the U.S. Army Corps of Engineers will generally not occur until state and local requirements are met. Since the Maryland and Virginia permitting procedures differ in some ways they are discussed separately:

Maryland Wetlands Act

Maryland's Wetlands Act of 1970 superseded earlier laws in making some important distinctions concerning the treatment of tidelands. Important changes were made regarding riparian rights, and provisions were made to regulate the use of wetlands. Since Maryland is a *"high-water state,"* meaning that the state owns the land up to mean high tide, distinctions were made in the regulatory treatment of intertidal (state) and supertidal (private) wetlands.

An individual planning a project that may affect private or state wetlands should contact the Wetlands Division, Water Resources Administration, Maryland Department of Natural Resources, Tawes State Office Building, Annapolis, MD 21401.

Virginia Wetlands Act

The 1972 Virginia Wetlands Act provides for the protection of vegetated wetlands up to one-and-a-half times the tidal range above mean low water. In 1982 the act was amended to include nonvegetated areas between mean high tide and mean low tide. Virginia is a *"low-water state,"* meaning that the area between mean high tide and mean low tide is privately owned. Thus wetlands-protection laws in Virginia deal primarily with private wetlands. Virginia established the option of enforcement through local wetlands boards or through the Marine Resources Commission. Most communities opted to establish local boards; the enforcement power in an area of interest can be found by contacting the Marine Resources Commission, Environmental Affairs Division, P.O. Box 756, Newport News, VA 23607.

To help simplify the process of obtaining all of the necessary local, state, and federal permits, Virginia, working in conjunction with the U.S. Army Corps of Engineers, has devised a joint permit application. This is designed to eliminate duplication of effort of both the individual and the regulatory agencies.

Whether a project will run into difficulty meeting permitting requirements depends on many factors including its scope, the type of construction involved, methods used, local environmental characteristics, and specific local and state laws. For example, dredge-and-fill projects are potentially more disturbing to wetlands ecosystems than are open piling structures such as piers, and are thus more regulated.

Technical assistance in understanding the regulations and permitting procedures, and in determining how best to plan projects for sensitive areas can be obtained from the state agencies mentioned above and from the appropriate district office of the U.S. Army Corps of Engineers listed in appendix B.

7 Building or buying a house near the shore

Life's important decisions are based on an evaluation of the facts. Few of us buy goods, choose a career, take legal, financial, or medical actions without first evaluating the facts and seeking advice. In the case of coastal property, two general areas should be considered: site safety, and the integrity of the existing or proposed building relative to the forces to which it will be subjected. Chapters 4 and 5 provide a guide to evaluating the site of your interest. Even the lowest risk site, however, may not provide security against storm or flood if a dwelling is not properly constructed.

Coastal realty versus coastal reality

Coastal property is not the same as inland property. Do not approach it as if you were buying a lot in a developed part of the mainland or a subdivided farm field in the country. The previous chapters illustrate that the shores of Virginia and Maryland, especially the barrier islands, capes, spits, and bluffs, are composed of variable environments and are subject to nature's most powerful and persistent forces. The reality of the coast is its dynamic character. Property lines are an artificial grid superimposed on this dynamism. If you choose to place yourself or others in this zone, prudence is in order.

A quick glance at the architecture of coastal buildings provides convincing evidence that the reality of coastal processes was rarely considered in their construction. Not too many years back, old-timers wisely lived back from the shore, behind the protection of sand dunes if they were present. More recently, newcomers to the shore have built precariously close to the bluff edge, with footings on spongy marsh subsoil, or atop of dunes as if to get a better view of future storms.

Many structures along the coastline were built before the adoption of a standardized building code or Federal Insurance Administration (FIA) guidelines. Town building officials enforced whatever codes were used, hence there is a wide disparity in the quality of these structures. New construction and major renovation projects must adhere to the FIA guidelines, but many existing dwellings were built before these requirements became effective.

This chapter provides an introduction to building near the shore and tells what to look for in an existing house or building. Emphasis is on the structure itself, whether a cottage, condominium, or mobile home. Keep in mind, however, that this is a general summary—application varies with region, type and intensity of hazards, type of shore material, and even social and political factors (e.g., density of development, quality of adjacent structures, and enforcement of building codes and zoning ordinances). A structure in the upper Bay, far removed from open ocean and not on a barrier-island shore, probably will survive longer with less stringent building requirements than a similar building in the lower Bay or on the Virginia barrier coast. This is not to say that persons in the upper Bay shouldn't follow prudent construction guidelines.

For those who want to learn more about construction near the shore, we recommend the book, *Coastal Design: A Guide for Builders, Planners, and Homeowners* (reference 138, appendix C), which gives more detail on coastal construction and may be used to supplement this volume. In addition, the Federal Emergency Management Agency's *Coastal Construction Manual* (reference 139, appendix C) is a good source of information on coastal construction and contains additional reference material. Persons considering building in areas threatened by both flooding and wave energy, especially those on barrier islands or low areas adjacent to the Bay, where storm surge is likely, should obtain *Elevated Residential Structures* (reference 140, appendix C) from FEMA. Other useful references are listed in appendix C.

The structure: concept of balanced risk

The constraints of economy and environment create a chance of failure for any structure. The objective of building design is to create a structure that is both economically feasible and functionally reliable. A house must be affordable and have a reasonable life expectancy, free of being damaged, destroyed, or wearing out. In order to obtain such a house, a balance must be achieved among financial, structural, environmental, and other special conditions. The effects of most of these conditions are heightened on the coast —property values are higher, there is a greater desire for aesthetics, the environment is more sensitive, the likelihood of severe storms is increased, and there are more hazards with which to deal.

The individual who builds or buys a house in an area exposed to erosion and flooding hazards or storm damage should understand the risks and weigh them against the benefits to be derived from living at this location. Similarly, the developer who is putting up a motel should weigh the possibility of damage or even destruction during a hurricane versus the money and other advantages to be gained from such a building. Then and only then should construction proceed. For both the homeowner and the developer, proper construction and location reduce the risks.

The concept of balanced risk should take into account the following fundamental considerations:

1. A coastal structure, because it is exposed to high winds,

waves, or flooding, should be built stronger than a structure built inland.
2. A building that houses elderly or sick people should be built safer than a building housing able-bodied people.
3. Because construction must be economically feasible, ultimate and total safety is not obtainable for most homeowners on the open-ocean coast. However, structures can be designed and built to resist all but the largest storms and still be within reasonable economic limits.
4. A building with a planned long life, such as a year-round residence, should be stronger than a building with a planned short life, such as a mobile home or a summer cottage.

Structural engineering is the designing and constructing of buildings to withstand the forces of nature. It is based on a knowledge of the forces to which the structures will be subjected and an understanding of the strength of building materials. The effectiveness of structural-engineering design was reflected in the aftermath of Typhoon Tracy, which struck Darwin, Australia, in 1974: 70 percent of housing that was not based on structural-engineering principles was destroyed and 20 percent was seriously damaged. Only 10 percent of the nonstructurally engineered housing weathered the storm. In contrast, over 70 percent of the structurally engineered, large commercial, government, and industrial buildings came through with little or no damage, and less than 5 percent of such structures were destroyed. Housing accounts for more than half of the capital cost of all the buildings in Queensland, and after the storm the state government established a building code requiring standardized structural engineering for houses in storm-prone areas. This improvement has been achieved with little increase in construction and design costs.

Coastal forces: design requirements

Wind

Hurricanes, with their associated high winds and storm surge topped by large waves, are the most destructive of the forces to be reckoned with on the coast. However, the frequency of hurricanes is much less than that of nor'easters, so the total damage of nor'easters is greater than that of hurricanes.

Winds can be evaluated in terms of the pressure they exert. Wind pressure varies with the square of the velocity, the height above the ground, and the shape of the object against which the wind is blowing. For example, a 100 mph wind exerts a pressure or force of about 40 pounds per square foot (psf) on a flat surface such as a sign. The effect on a curved surface such as a sphere or cylinder is much less—about one-half that on the flat surface. If the 100-mph wind picks up to 190 mph velocity, the 40 psf force would increase to 140 psf on a flat surface.

Wind velocity increases with height above ground, so a tall structure is subject to greater velocity and thereby greater pressure than a low structure. The wind velocity and corresponding pressure could be almost double at 100 feet above the ground than that at ground level.

A house or building designed for inland areas is built primarily to resist vertical loads. It is assumed that the foundation and framing must support the load of the walls, floor, roof, and furniture with relatively insignificant wind forces.

A well-built house in a hurricane-prone area, however, must be constructed to withstand a variety of strong wind forces that may come from any direction. Although many people think that wind damage is caused by uniform horizontal pressures (lateral loads), most damage, in fact, is caused by uplift (vertical), suctional (pressure outward from the house), and twisting (torsional) forces (figs. 7.1 and 7.2). High horizontal pressure on the windward side is accompanied by suction on the leeward side. The roof is subjected both to downward pressure and to uplift. Often a roof is sucked up by the uplift drag of the wind. Usually the failure of houses is in the devices that tie the parts of the structure together. All structural members (beams, rafters, columns) should be fastened together on the assumption that about 25 percent of the vertical load on the member may be a force coming from any direction (sideways or upward). Such structural integrity is also important if it is likely that the structure may have to be moved to avoid destruction by shoreline retreat.

Storm surge and flooding

Storm surge is a rise in sea level above the normal water level during a storm. During hurricanes and other storms, the coastal zone is often inundated by storm surges and accompanying storm waves, and these cause most property damage and loss of life.

Often the wind backs water into streams or estuaries already swollen from the exceptional rainfall brought on by the hurricane or nor'easter. Water is piled onto the shore by the storm. In some cases the direction of flooding may be from the bay or landward side of coastal islands. This flooding is particularly dangerous when the wind pressure keeps the tide from running out of inlets, so that the next normal high tide pushes the accumulated waters even higher. Flooding can cause an unanchored house to float off its foundation and come to rest against another house, severely damaging both. Even if the house itself is left structurally intact, flooding may destroy its contents. People who have cleaned the mud and contents out of a flooded house retain vivid memories of the effects.

Proper coastal development takes into account the expected level and frequency of storm surge for the area. In general, building standards require that the first habitable floor of a dwelling be above the 100-year–flood level plus an allowance for wave height. At this level, a building has a 1 percent probability of being flooded in any given year.

170 Building or buying a house

WIND

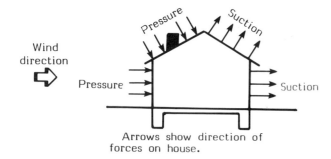

Arrows show direction of forces on house.

WAVES

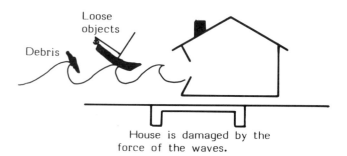

House is damaged by the force of the waves.

DROP IN BAROMETRIC PRESSURE

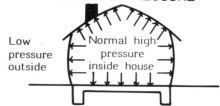

The passing eye of the storm creates different pressure inside and out, and high pressure inside attempts to burst house open.

HIGH WATER

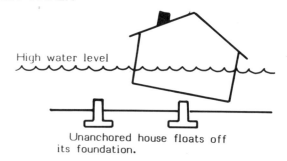

Unanchored house floats off its foundation.

7.1 Forces to be reckoned with at the shore.

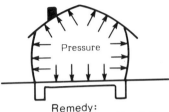
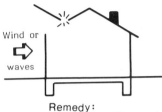
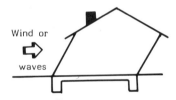
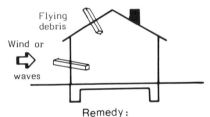

7.2 Modes of failure and how to deal with them. Modified from *Wind Resistant Design Concepts for Residences* (reference 146, appendix C).

Waves

Hurricane and persistent storm waves can cause severe damage not only in forcing water onshore to flood buildings, but also in throwing boats, barges, piers, houses, and other floating debris inland against standing structures. To understand the force of a wave, consider that a cubic yard of water weighs over three-fourths of a ton; hence, a breaking wave moving shoreward at a speed of several tens of miles per hour can be one of the most destructive elements of a hurricane. Waves can also destroy coastal structures by scouring away the underlying sand, which causes the structures to collapse. It is possible to design buildings to survive crashing storm surf as in the case of lighthouses but in the balanced-risk equation it usually is not economically feasible. On the other hand, even cottages can be made considerably more storm worthy by following the suggestions that follow in this chapter.

Battering by debris

Even though it may be an isolated occurrence, the likelihood of floating objects striking a house during flooding should be taken into account. To get an idea of the size of a battering load against which the house should be designed, we refer to the *Model Minimum Hurricane Resistant Building Standards for the Texas Gulf Coast* (reference 161, appendix C), which specifies that a structure should withstand a battering load equal to the impact force produced by a 1,000-pound mass traveling at a velocity of 10 feet per second and acting on a one-square-foot surface.

The above standards specify that certain buildings such as "safe refuges" be constructed to resist more severe battering loads than the above-listed normal load. A *safe refuge* is a building or structure located in the flood area with space sufficiently above the high water level to be authorized as a haven in the event of a hurricane.

Barometric-pressure change

Changes in barometric pressure may be a minor contributor to structural failure. If a house is sealed at a normal barometric pressure of 30 inches of mercury, and the external pressure suddenly drops to 26.61 inches of mercury (as it did in Hurricane Camille in Mississippi in 1969), the pressure exerted within the house is 245 pounds psf. If a house were leakproof, it would explode. Fortunately, houses leak air, so given the most destructive forces of storm wind and waves, pressure differential is of minor concern. Venting the underside of the roof at the eaves is a common means of equalizing internal and external pressure.

House selection

Some types of houses are more suited than others for the shore, and knowing the differences will help you make a better selection, whether you are building a new house or buying an existing one.

It is hard to beat a wood-frame house that is properly braced and anchored and has well-connected members. The well-built wood house will often hold together as a unit even if it is moved off its foundation. A wood-frame house is also the easiest to raise to a safer level, a common flood-proofing technique for older houses. However, even a wood-frame building must be designed (or modified) and adequately anchored to prevent flotation collapse or lateral movement. It must be constructed with materials and utility equipment resistant to flood damage.

Unreinforced masonry houses, whether they be brick, concrete block, hollow clay tile, or brick veneer, are not as able as solid wood-frame houses to withstand the lateral forces of wind and waves, the battering of debris, the flooding, the scour, and the settling of the foundation. Extraordinary reinforcing in coastal regions will alleviate the inherent weakness of unit masonry, if done properly. Reinforced concrete and steel frames are excellent, but are rarely used in constructing small residential structures. Another consideration is that masonry houses are difficult to move. Shorelines do retreat, and it may be more economical to move a movable structure than to protect an immovable house.

Keeping dry: pole or stilt houses

In coastal regions subject to flooding by waves or storm surge, the best and most common method of minimizing damage is to raise the lowest floor of a residence above the expected water level. The first habitable floor of a home must be above the 100-year–storm-surge level (plus calculated wave height) to qualify for NFIP.

Nonresidential buildings should be flood proofed at least up to the base flood level or elevated at or above this level. As a result, where the soil is suitable, most modern flood-zone structures should be constructed on piling, well anchored in the subsoil. Elevating the structure by building a mound is adequate if flooding is the only hazard, but it is not suited to the coastal zone, where mounded soil is easily eroded.

Current building-design criteria for pole-house construction under NFIP are outlined in the book *Elevated Residential Structures* (reference 140, appendix C). Regardless of insurance, pole-type construction with deep embedment of the poles is best in areas where waves and storm surge will erode foundation material (e.g., barrier islands, dune and marsh areas). Materials used in pole construction include the following:

Piles. Long, slender columns of wood, steel, or concrete driven into the earth to a sufficient depth to support the vertical load of the house and to withstand horizontal forces of flowing water, wind, and water-borne debris. Pile construction is especially suitable in areas where scouring (soil washing out from under the foundation of a house) is a problem.

Posts. Usually composed of wood; if of steel, they are called "columns." Unlike piles, they are not driven into the ground, but are placed in a pre-dug hole at the bottom of which may be a concrete pad (fig. 7.3). Posts may be held in place by backfilling and tamping earth, or by pouring concrete into the hole after the post is in place. Posts are more readily aligned than driven piles

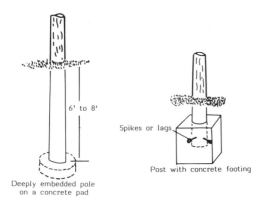

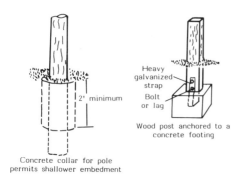

7.3 Shallow and deep supports for poles and posts. Source: Southern Pine Association.

and therefore are better to use if poles must extend to the roof. In general, treated wood is the cheapest and most common material for both posts and piles.

Piers. Vertical supports, thicker than piles or posts, usually made of reinforced concrete or reinforced masonry (concrete blocks or bricks). They are set on footings and extend to the underside of the floor frame.

Pole construction (fig. 7.4) can be of two types. The poles can be cut off at the first-floor level to support a platform that serves as the dwelling floor. In this case, piles, posts, or piers can be used. Or they can be extended to the roof and rigidly tied into both the floor and the roof. In this way, they become major framing members for the structure and provide better anchorage to the house as a whole (fig. 7.5). A combination of full and floor-height poles is used in some cases, with the shorter poles restricted to supporting the floor inside the house (fig. 7.6).

Where the foundation material can be eroded by waves or winds, the poles should be deeply embedded and solidly anchored either by driving piles or by drilling deep holes for posts and putting in a concrete pad at the bottom of each hole. If the embedment is shallow, a concrete collar around the poles improves anchorage. The choice depends on the soil conditions. In either case, the foundations must be deep enough to provide support after maximum predicted loss of sand from storm erosion and scour. This required depth will often dictate piles rather than posts.

Piles permit far better penetration than posts. This is important because a structure can fail if storm waves liquefy and erode sand support and the depth of piles or posts is not sufficient. Improper connections of floor to piling and inadequate pile bracing can also cause structural failure. Just as important as driving the piling deep enough to resist scouring and to support the loads they must carry is the need to fasten them securely to the structure above them. The connections must resist horizontal loads from wind and wave during a storm and also uplift from the same source.

Some localities require that piles be driven to a depth of at least 10 feet below mean sea level. The floor and the roof should be securely connected to the poles with bolts or other fasteners. When poles do not extend to the roof, attachment is even more critical. A system of metal straps is often used. Unfortunately, it sometimes happens that builders inadequately attach the girders, beams, and joists to the supporting poles by too few and undersized bolts. Hurricanes have proven this to be insufficient.

Local building codes may specify the size, quality, and spacing of the piles, ties, and bracing, as well as the methods of fastening them to the structure. Building codes often are minimal requirements, however, and building inspectors are usually amenable to allowing designs that are equally or more effective.

If post holes are dug, rather than pilings driven, the posts should extend 4 to 8 feet into the ground to provide anchorage. As previously stated, the lower end of the post should rest on a concrete pad, spreading the load to the soil over a greater area to prevent settlement. Where the soil is sandy or is the type that the embedment can be less than approximately 6 feet or so, it is best to tie the post down to the footing with straps or other anchoring devices

7.4 (A) Pole house construction at Sandbridge, Virginia. (B) Poles in place prior to construction at Virginia Beach.

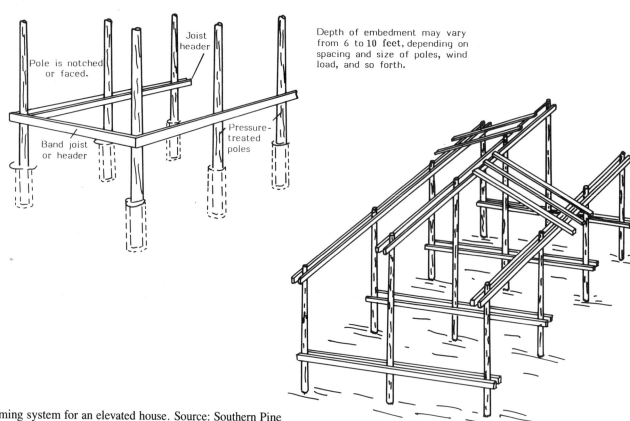

7.5 Framing system for an elevated house. Source: Southern Pine Association.

Building or buying a house **177**

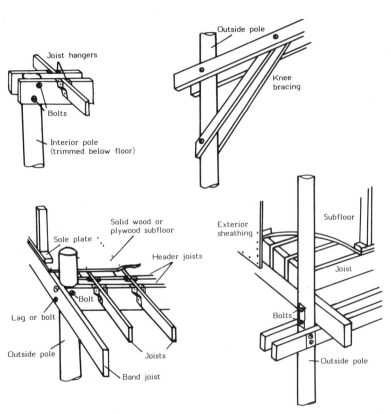

to prevent uplift. Driven piles should have a minimum penetration of 8 feet. However, most soils require greater embedment—especially if the site is near the water.

The space under an elevated house, whether pole type or otherwise, must be kept free of obstructions in order to minimize the impact of waves and floating debris. If the space is enclosed, the enclosing walls should be designed so that they can break away or fall under flood loads, but they must remain attached to the house or be heavy enough to sink. Otherwise, the walls will float away and add to the water-borne debris problem. Ways of avoiding this problem are designing walls that can be swung up out of the path of the floodwaters, or building them with louvers that will allow water to pass through. The louvered wall is subject to damage from floating debris. It is tempting to close in the ground floor for a garage, storage, or recreation room, but the practice may be costly. It may violate insurance requirements and actually contribute to the loss of the house in a hurricane. The design of enclosing breakaway walls should be checked against insurance requirements. See *Elevated Residential Structures* (reference 140, appendix C).

7.6 Tying floors to poles. Source: Southern Pine Association.

An existing house: what to look for

If instead of building a new house, you are selecting a house already built in an area subject to waves, flooding, or high winds, consider the following factors: (1) where the house is located; (2) how well the house is built; and (3) how the house can be improved.

Geographic location

Evaluate the site of an existing house using the same principles given in earlier chapters for the evaluation of a possible site to build a new house. Among other things, consider erosion rates, flood elevations, house elevation, frequency of high water, escape route, drainage, and movability.

You can modify the house after you have purchased it, but you can't prevent hurricanes or other storms. The first step is to stop and consider: Do the pleasure and benefits of this location balance the risk and disadvantages? If not, look elsewhere for a home; if so, then evaluate the house itself.

Quality of construction

In general, the principles used to evaluate an existing house are the same as those used in building a new one. It should be remembered that many houses predate the enactment of the National Flood Insurance Program and may not meet the standards or improvements required of structures built since then.

Before you thoroughly inspect the building in which you are interested, look closely at the adjacent structures. If poorly built, they may float over against your building and damage it in a flood. You may even want to consider the type of people you will have as neighbors: Will they "clear the decks" in preparation for a storm or will they leave items in the yard to become wind-borne missiles?

Next, inspect the house or condominium itself for anchorage. If it is simply resting on blocks, rising water may cause it to float off its foundation and come to rest against your neighbor's house or out in the middle of the street. If an unanchored house is well built and well braced internally, it may be possible to move it back to its proper location, but chances are great that it will be too damaged to be habitable.

If the building is on piles, posts, or poles, check to see if the floor beams are adequately bolted to them. If it rests on piers, crawl under the house if space permits to see if the floor beams are securely connected to the foundation. If the floor system rests unanchored on piers, it will be more likely to be moved by storm forces.

It is more difficult to discern whether a building built on a concrete slab is properly bolted to the slab because the inside and outside walls hide the bolts. If you can locate the builder, ask if such bolting was done. Better yet, get assurance that construction of the house complied with the provisions of a building code serving the needs of that particular region. With such assurance, you can be reasonably sure that all parts of the house are well anchored: the foundation to the ground; the floor to the foundation; the walls to the floor; and the roof to the walls (figs. 7.7, 7.8, and 7.9). Be aware, however, that some builders, carpenters, and building inspectors who are accustomed to traditional construction are apt to regard metal connectors, collar beams, and other such devices as newfangled and unnecessary. If consulted, they may assure you that a house is as solid as a rock, when in fact it is far from it. Nevertheless, it is wise to consult the builder or knowledgeable neighbors when possible.

The roof should be well anchored to the walls. This will prevent uplifting and separation. Visit the attic to see if such anchoring exists. Simple toenailing (nailing at an angle) is not adequate; metal fasteners are needed. Depending on the type of construction and the amount of insulation laid on the floor of the attic, these may or may not be easy to see. If roof trusses or braced rafters were used, it should be easy to see whether the various members, such as the diagonals, are well fastened together. Again, simple toenailing will not suffice. Some builders, unfortunately, nail parts of a roof truss just enough to hold it together to get it in place. A *collar beam* or *gusset* at the peak of the roof (fig. 7.10) provides some assurance of good construction. The Standard Building Code states that wood truss rafters must be securely fastened to the exterior walls with approved hurricane anchors or clips.

Quality roofing material should be well anchored to the sheath-

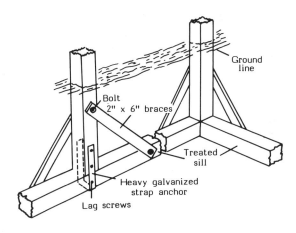

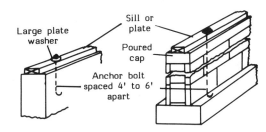

7.7 Foundation anchorage. Top: anchored sill for shallow embedment. Bottom: anchoring sill or plate to foundation. Bottom modified from *Houses Can Resist Hurricanes* (reference 157, appendix C).

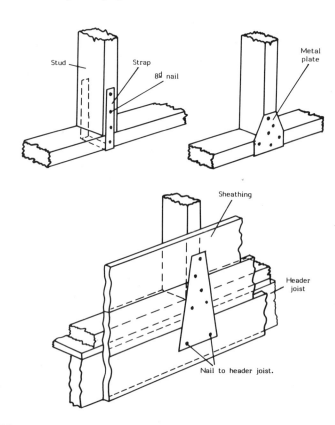

7.8 Stud-to-floor, plate-to-floor framing methods. Modified from *Houses Can Resist Hurricanes* (reference 157, appendix C).

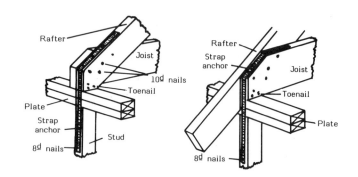

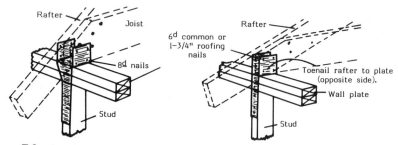

7.9 Roof-to-wall connectors. The top drawings show metal strap connectors. Left, rafter to stud; right, joist to stud. The bottom left drawing shows a double-member metal plate connector—in this case with the joist to the right of the rafter. The bottom right drawing shows a single-member metal plate connector. Modified from *Houses Can Resist Hurricanes* (reference 157, appendix C).

ing. A poor roof covering will be destroyed by hurricane-force winds, allowing rain to enter the house and damage ceilings, walls, and the contents of the house. Galvanized nails (two per shingle) should be used to connect wood shingles and shakes to wood sheathing, and they should be long enough to penetrate through the sheathing. (*Sheathing* is the covering [usually wood boards, plywood, or wall boards] placed over rafters or the exterior studding of a building to provide a base for the application of roof or wall cladding.) Threaded nails should be used for plywood sheathing.

Also consider how much of the shingle is exposed. For roof slopes that rise one foot for every three feet or more of horizontal distance, exposure of the shingle should be about one-fourth of its length (four inches for a 16-inch shingle). If shakes (thicker and longer than shingles) are used, less than one-third of their length should be exposed. In hurricane areas, asphalt shingles should be exposed somewhat less than usual. A mastic or seal-tab type, or an interlocking shingle of heavy grade should be used along with a roof underlay of asphalt-saturated felt. Galvanized roofing nails or approved staples are required.

The fundamental rule to remember in framing is that all structural elements should be fastened together and anchored to the ground in such a manner as to resist all forces, regardless of which direction these forces may come from. This prevents overturning, floating off, racking, or disintegration.

The shape of the house is also important. A *hip roof*, which slopes in four directions, is better able to resist high winds than a *gable roof*, which slopes in two directions. This was found to be true in Hurricane Camille in 1969 in Mississippi, and, later, in Typhoon Tracy, which devastated Darwin, Australia, in December, 1974. The reason is twofold: the hip roof offers a smaller shape for the wind to blow against, and its structure is such that it is better braced in all directions.

Also note the horizontal cross section of the house (the shape of the house as viewed from above). The pressure exerted by a wind on a round or elliptical shape is about 60 percent of that exerted on the common square or rectangular shape; the pressure exerted on a hexagonal or octagonal cross section is about 80 percent of that exerted on a square or rectangular cross section.

The design of a house or building in a coastal area should minimize structural discontinuities and irregularities. It should be plain and simple and have a minimum of nooks and crannies and offsets on the exterior, because damage to a structure tends to concentrate at these points. Some of the newer beach-cottage designs are highly angular with many such nooks and crannies. Award-winning architecture will be a storm loser if the design has not incorporated the technology for maximizing structural integrity with respect to storm forces. Simple, regular construction allows the house to react to storm winds as a complete unit.

Brick, concrete-block, and masonry-wall houses should be adequately reinforced. This reinforcement is hidden from view. Building codes applicable to high-wind areas often specify the type of mortar, reinforcing, and anchoring to be used in construction.

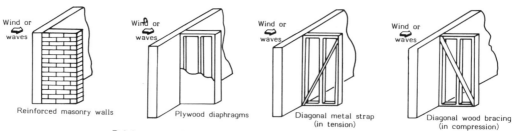

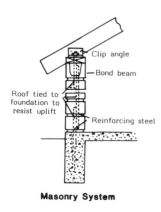

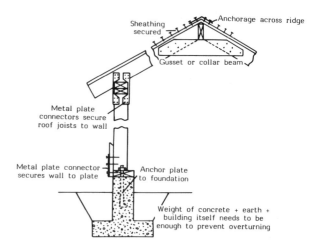

7.10 Where to strengthen a house. Modified from *Wind Resistant Design Concepts for Residences* (reference 146, appendix C).

If you can get assurance that the house was built in compliance with a building code designed for such an area, it may be a good investment. Avoid unreinforced masonry houses.

A poured-concrete *bond beam* at the top of the wall just under the roof is one indication that the house is well built (fig. 7.11). Most bond beams are formed by putting in reinforcing and pouring concrete in U-shaped concrete blocks. From the outside, however, you can't distinguish these U-shaped blocks from ordinary ones and therefore can't be certain that a bond beam exists. The vertical reinforcing should penetrate the bond beam.

Because they think it looks better some architects and builders use a *stacked bond* (one block directly above another), rather than overlapped or staggered blocks. The stacked bond is definitely weaker. Unless you have proof that the walls are reinforced adequately to overcome this lack of strength, you should avoid this type of construction.

Glazing (windows, glass doors, glass panels) should be minimal. Although large, open glass areas facing open water provide an excellent bay view or sea view, such glazing may present several problems. The obvious hazard is disintegrating and inward-blowing glass during a storm. Glass projectiles are lethal. Less frequently recognized problems include the fact that glass lacks the structural strength of wood, metal, or other building materials; and ocean-facing glass is commonly damaged through sand blasting, transported by normal coastal winds. The solution may be to reduce the amount of glass in the original design, or install storm shutters, which come in a variety of materials from steel to wood.

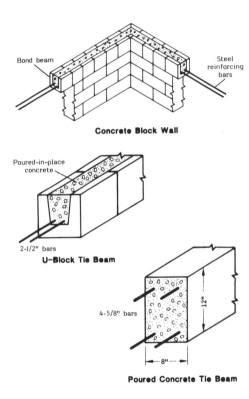

7.11 Reinforced tie beam (bond beam) for concrete block walls—to be used at each floor level and at roof level around the perimeter of the exterior walls.

Consult a good architect or structural engineer for advice if you are in doubt about any aspect of a house. A few dollars spent for wise counsel may save you from later financial grief.

To summarize, a shore house should have: (1) roof tied to walls, walls tied to foundation, and foundation anchored to the earth; (2) a shape that resists storm forces; (3) floors high enough to be above most storm waters (usually the 100-year–flood level plus 3 to 8 feet); (4) piles or posts of sufficient depth or embedded in concrete to anchor the structure and withstand erosion; and (5) well-braced piling (fig. 7.12).

Improvement possibilities

If you currently own a house or are contemplating buying one in a hurricane-prone area, you will want to know how to protect both the house and its occupants. As a first step, obtain the excellent publication, *Wind Resistant Design Concepts for Residences*, by Delbart B. Ward (reference 146, appendix C). Of particular interest are the sections on building a refuge-shelter module within a residence. Also noteworthy are two supplements to this publication (reference 147, appendix C), which deal with buildings larger than single-family residences in urban areas.

Suppose your house is resting on blocks but is not fastened to them and, thus, is not adequately anchored to the ground. Can anything be done? One solution is to treat the house like a mobile home. Screw ground anchors into the ground to a depth of four feet or more and fasten them to the underside of the floor system.

Calculations to determine the needed number of ground anchors will differ between a house and a mobile home, because each is affected differently by the forces of wind and water. See figs. 7.13 and 7.14 for illustrations of how ground anchors can be used.

Recent practice is to put commercial steel-rod anchors in at an angle in order to better align them with the direction of the pull. If a vertical anchor is used, the top 18 inches or so should be encased in a concrete cylinder about 12 inches in diameter. This prevents the top of the anchor rod from bending or slicing through the wet soil from the horizontal component of the pull.

Diagonal struts, either timber or pipe, may be used to anchor a house that rests on blocks. This is done by fastening the upper ends of the struts to the floor system, and the lower ends to individual concrete footings substantially below the surface of the ground. These struts must be able to take both uplift (tension) and compression, and should be tied into the concrete footing with anchoring devices such as straps or spikes.

If the house has a porch with exposed columns or posts, it should be possible to install tie-down anchors on their tops and bottoms. In most cases, steel straps should suffice.

When accessible, roof rafters and trusses should be anchored to the wall system. Usually the roof trusses or braced rafters are sufficiently exposed to make it possible to strengthen joints (where two or more members meet) with collar beams or gussets, particularly at the peak of the roof (fig. 7.10).

The purpose of straps is to prevent any or all of the house from blowing away. The Standard Building Code says: "Lateral support

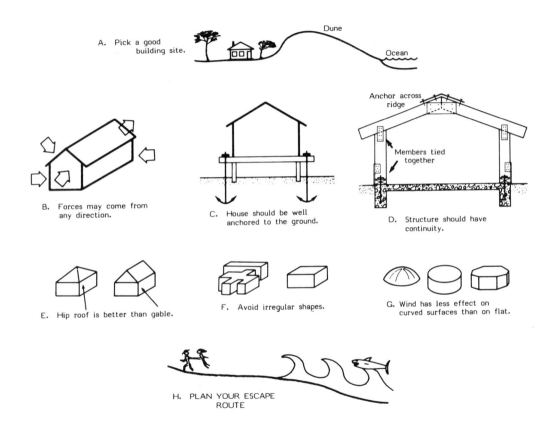

7.12 Some rules in selecting or designing a house.

securely anchored to all walls provides the best and only sound structural stability against horizontal thrusts, such as winds of exceptional velocity." The cost of connecting all elements securely adds very little to the cost of the frame of the dwelling, usually under 10 percent, and a very much smaller percentage of the total cost of the house.

If the house has an overhanging eave and there are no openings on its underside, it may be feasible to cut openings and screen them. These openings keep the attic cooler (a plus in the summer) and help to equalize the pressure inside and outside of the house during a storm with a low-pressure center.

Consider bracing or strengthening the interior walls. Such reinforcement may require removing the surface covering and installing plywood sheathing or strap bracing. Where wall studs are exposed, bracing straps offer a simple way to achieve needed reinforcement against the wind. These straps are commercially produced and are made of 16-gauge galvanized metal with pre-punched holes for nailing. These should be secured to studs and wall plates as nail holes permit (fig. 7.10). Bear in mind that they are good only for tension.

Another way a house can be improved is to modify one room so that it can be used as an emergency refuge in case you are trapped in a major storm. This suggestion is *not* an alternative to evacuation prior to a hurricane. Examine the house and select the best room to stay in during a storm. A small windowless room such as a bathroom, utility room, or storage space is usually stronger than a room with windows. A sturdy inner room, with more than one wall between it and the outside, is safest. The fewer doors, the better; an adjoining wall or baffle wall shielding the door adds to the protection.

A competent carpenter, architect, or structural engineer can review the house with you and help you decide what modifications are most practical and effective. Do not be misled by someone who resists new ideas.

Mobile homes: limiting their mobility

Nearly six million Americans live in mobile homes today and the number is growing. Twenty to thirty percent of single-family housing production in the United States consists of mobile homes. But because of their light weight and flat sides, mobile homes are vulnerable to the high winds of hurricanes, tornadoes, and severe storms. Such winds can overturn unanchored mobile homes or smash them into neighboring homes and property. High winds damage or destroy nearly five thousand mobile homes every year, and the number will surely rise unless protective measures are taken. As one man whose mobile home was overturned in Hurricane Frederic (1979) so aptly put it: "People who live in flimsy houses shouldn't have hurricanes."

Several lessons can be learned from past storm experiences. First, mobile homes should be located properly. After Hurricane Camille (1969), it was observed that damage was minimized where mobile-home parks were surrounded by woods and the units were close together. The damage that did occur was caused mainly by fallen trees. In unprotected areas, however, many mobile homes were overturned and often destroyed from the force of the wind.

A hilltop location will greatly increase a mobile home's vulnerability to the wind. A lower site screened by trees is safer from the wind, but it should be above storm-surge flood levels. A location that is too low is obviously more likely to flood. There are fewer safe locations for mobile homes than for stilt houses, although mobile homes can be elevated on pilings similar to pole or stilt houses.

Another lesson taught by past experience is that the mobile home must be tied down or anchored to the ground so that it will not overturn in high winds (figs. 7.13 and 7.14, table 7.1). Simple prudence dictates the use of tiedowns. Many insurance companies, moreover, will not insure mobile homes unless they are adequately anchored with tiedowns. A mobile home may be tied down with cable or rope, or may be rigidly attached to the ground by connecting it to a simple wood-post foundation system. In general, an entire tiedown system costs only a nominal amount.

A mobile home should be properly anchored with both ties to the frame and over-the-top straps. Otherwise, it may be damaged by sliding, overturning, or tossing. The most common cause of major damage is the tearing away of most or all of the roof. When this happens the walls are no longer adequately supported at the top, and are prone to collapse. Total destruction of a mobile home is likely if the roof blows off, especially if the roof blows off first and then the home overturns. The necessity for anchoring cannot be overemphasized; there should be over-the-top tiedowns to resist overturning, and frame ties to resist sliding off the piers. This applies to single mobile homes up to 14 feet in width. Double wides do not require over-the-top ties, but they do require frame ties. Although newer mobile homes are equipped with built-in straps to aid in tying down, the occupant may wish to add more if in a particularly vulnerable location. Many of the older homes are not equipped with built-in straps.

For more information, obtain a copy of *Manufactured Home Installation in Flood Hazard Areas* (reference 153, appendix C) from FEMA.

High-rise buildings: the urban shore

A high-rise building on the beach is generally designed by an architect and a structural engineer who are presumably well qualified and aware of the requirements for building on the shoreline. Tenants of such a building, however, should not assume that it is invulnerable. People living in apartment buildings of two or three stories were killed when the buildings were destroyed by Hurricane Camille. Storms in Delaware have smashed five-story buildings. Larger high rises have yet to be thoroughly tested by a major hurricane.

The first aspect of high-rise construction that a prospective dweller or condo owner must consider is the quality of the foundation. High rises near the beach should be built so that even if the foundation is severely undercut during a storm the building will remain standing. Pilings must be driven deeply enough to resist

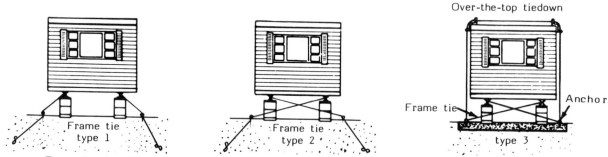

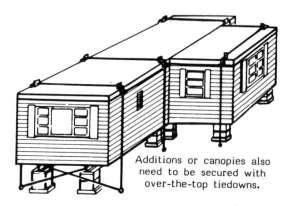

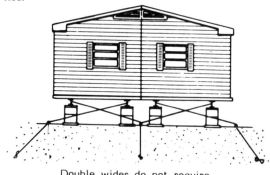

7.13 Tiedowns for mobile homes. Modified from *Protecting Mobile Homes from High Winds* (reference 154, appendix C).

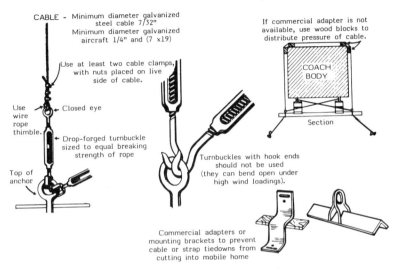

7.14 Hardware for mobile home tiedowns. Modified from *Protecting Mobile Homes from High Winds* (reference 154, appendix C).

scouring and to support the load they must carry. Just as important is the need to fasten piles securely to the structure. The connections must resist horizontal loads and uplift from wind and waves during a storm. Both builders and building inspectors must ensure that the job is done right.

Pressure from the wind is greater near the shore than it is inland, and it increases with height. If you are living inland in a two-story house and move to the eleventh floor of a high rise on the shore, you should expect much more wind pressure. This high wind pressure can cause unpleasant motion of the building. It is worthwhile to check with current residents of a high rise to find out if it has undesirable motion characteristics. More seriously, high winds can break windows, damage property, and inflict injury.

Those who are interested in researching the subject further—even the knowledgeable engineer or architect who is engaged to design a structure near the shore—should obtain a copy of *Structural Failures: Modes, Causes, Responsibilities* (reference 155, appendix C). Of particular importance is the chapter entitled, "Failure of Structures Due to Extreme Winds." This chapter analyzes wind damage to engineered high-rise buildings from the storms at Lubbock and Corpus Christi, Texas, in 1970.

Another concern in a multifamily high-rise building is the risk of power failure or blackout. Such an occurrence is more likely along the coast than inland because of the more severe weather conditions associated with coastal storms. A power failure can cause great distress, and its inconveniences and dangers tend to be more serious in a multifamily, multistory high rise than in a smaller low-occupancy structure.

Modular-unit construction: prefabricating the urban shore

An increasingly popular method of shoreline development is the practice of building a house, duplex, or large condominium structure by fabricating modular units in a shop and assembling them at the site. The larger of these structures are commonly two or three stories in height, and may contain a large number of living units.

Modular construction makes good economic sense, and there is nothing inherently wrong in this approach to coastal construction. These methods have been used in the manufacturing of mobile homes for years, although final assembly of mobile homes occurs in the shop rather than in the field. Doing as much of the work as possible in a shop can save considerable labor and cost. The workers are not affected by outside weather conditions. They often can be paid with piecework rates, enhancing their productivity. Shop work lends itself to labor-saving equipment such as pneumatic nailing guns and overhead cranes.

If the manufacturer desires it, shop fabrication can permit higher quality. Inspection and control of the whole process is much easier. For instance, there is less hesitation about rejecting a poor piece of lumber when you have a nearby supply of it than if you are building a single dwelling and have just so much lumber on the site.

On the other hand, because so much of the work is done out of the sight of the buyer, a manufacturer can take shortcuts if he is so inclined. Therefore, it is important to consider the following: Were wiring, plumbing, heating, air conditioning, and ventilation installed at the factory or at the building site? Were the installers licensed and certified? Was the work inspected at both the factory and on the construction site? Are the modular units properly fastened together to withstand coastal storms? It is vital that stacked units be adequately fastened together to resist high winds. An upper unit cannot depend solely on its weight to hold it in place.

In general, it is desirable to check the reputation and integrity of the manufacturer just as you would when hiring a contractor to build your individual house on site. Buying a modular unit should be approached with the same caution as for other structures.

In locating a modular unit as with all other types of structures, consider site safety and escape routes.

Table 7.1 Tiedown anchorage requirements

	10- and 12-foot-wide mobile homes				12- and 14-foot-wide mobile homes 60 to 70 feet long	
	30 to 50 feet long		50 to 60 feet long			
Wind velocity (mph)	Number of frame ties	Number of over-the-top ties	Number of frame ties	Number of over-the-top ties	Number of frame ties	Number of over-the-top ties
70	3	2	4	2	4	2
80	4	3	5	3	5	3
90	5	4	6	4	7	4
100	6	5	7	5	8	6
110	7	6	9	6	10	7

Living with nature: prudence pays

Hurricane or calm, receding shore or growing shore, storm-surge flood or sunny sky, migrating dune or maritime forest, win or lose, the gamble of coastal development will continue. You can, however, reduce the risks of shoreline living by choosing your site with natural safety in mind, following sound structural-engineering principles in construction, and taking a generally prudent approach to living and enjoying the shore.

Appendix A
Hurricanes

Ranking hurricanes: how bad is bad?

Hurricanes and other severe storms are always a threat in the Chesapeake Bay region. This is especially true in the open Atlantic and lower Bay coastal segments.

Hurricane chroniclers note that many historical accounts characterize each major storm as "the worst ever" or "greater than" the previous "worst" storm. Although storm activity may be cyclic, it is doubtful that storms have increased in intensity.

As coastal development has increased, storm damage has increased accordingly. We might conclude erroneously that Hurricane Betsy (1965) and Hurricane Camille (1969) were of equal strength because each storm caused damages totaling $1.4 billion. In reality, Hurricane Betsy was a weaker storm, but it struck more well-developed areas.

Similarly, loss of life cannot be used to measure storm intensity or to compare storms. Relatively small storms of a century ago were more deadly because they came without warning; there was no time to evacuate. This is why 6,000 people died in a hurricane at Galveston, Texas, in 1900. Today, advance warning, efficient evacuation, and safer construction results in lower casualty rates, even in a major hurricane. But unsafe development, allowing population growth to exceed the capacity for safe evacuation, and complacency on the part of coastal residents could reverse this trend with shocking results. The National Hurricane Center has warned repeatedly that tens of thousands of Americans could die if a major storm strikes certain low-elevation areas of heavy development such as in southern or western Florida. This is equally true for the heavily populated areas of the Virginia–Atlantic Coast.

To better warn coastal residents of the strength or intensity of an impending hurricane, the National Weather Service uses the Saffir-Simpson Scale (table A.1) to describe storms. The scale is based on three storm variables: wind velocity, storm surge, and barometric pressure. Considerable correlation exists between these variables, which when combined with a knowledge of the seabed and coastline of a given area can lead to more accurate and timely forecasting of hurricane impact.

Do not be misled by the scale, however. A hurricane is a hurricane, so the scale is defining how bad is bad. Regardless of whether the hurricane is a category 1 or category 5, when the word comes to evacuate, *do it*! Wind velocity may change or the configuration of the coast may amplify storm-surge level, so the category rank you hear in the news report may change by the time the storm reaches your position. Don't gamble with your life or the lives of others. *Go!*

Hurricane checklist

When a hurricane or severe storm is approaching and you have been advised to evacuate, it is difficult to determine what to bring, where to go, and what to do to your home to prepare for the imminent threat. Consequently, all coastal residents, especially those in the lower Bay or along the open Atlantic Coast, should prepare in advance. The remainder of this appendix provides a checklist to help prepare for a major storm ahead of time.

When a hurricane threatens

— Listen for official weather reports.
— Read your newspaper and listen to radio and television for official announcements.
— Note the address of the nearest emergency shelter.
— Know the official evacuation route in advance.
— Call a physician for advice if you are pregnant, ill, or infirm.
— Be prepared to turn off gas, water, and electricity where it enters your home.
— Fill tubs and containers with water (one quart per person per day).
— Make sure your car's gas tank is full.

Table A.1 Saffir-Simpson Hurricane Scale

Category	Winds (mph)	Storm surge (feet)	Central pressure (inches)	Damage
1	74–95	4–5	≥28.94	Minimal
2	96–110	6–8	28.50–28.91	Moderate
3	111–130	9–12	27.91–28.47	Extensive
4	131–155	13–18	27.17–27.88	Extreme
5	>155	>18	<27.17	Catastrophic

— Secure your boat. Use long lines to allow for rising water.
— Secure movable objects on your property:
  — doors and gates
  — outdoor furniture
  — shutters
  — garden tools
  — hoses
  — garbage cans
  — bicycles or large sports equipment
  — barbecues or grills
  — other
— Board up or tape windows and glassed areas. Close storm shutters. Draw drapes and window blinds across windows and glass doors. Remove furniture in their vicinity.
— Stock adequate supplies:
  — transistor radio
  — fresh batteries
  — canned heat
  — hammer
  — boards
  — flashlights
  — candles
  — matches
  — nails
  — screwdriver

- pliers
- hunting knife
- tape
- first-aid kit
- prescribed medicines
- water-purification tablets
- insect repellent
- gum, candy
- life jackets
- charcoal bucket and charcoal
- ax*
- rope*
- plastic drop cloths, waterproof bags, ties
- containers for water
- disinfectant
- canned food, juices, soft drinks (see below)
- hard-top headgear
- fire extinguisher
- can opener and utensils
- Check mobile-home tiedowns.

Example of storm food stock for family of four

- two 13-oz. cans evaporated milk
- four 7-oz. cans fruit juice
- two cans tuna, sardines, chicken
- three 10-oz. cans vegetable soup
- one small can of cocoa
- one 15-oz. box raisins or prunes
- salt
- pet food

*Take an ax (to cut an emergency escape opening) if you go to the upper floors or attic of your home. Take rope for escape to ground when water subsides.

- one 14-oz. can cream of wheat or oatmeal
- one 8-oz. jar peanut butter or cheese spread
- two 16-oz. cans pork and beans
- one 2-oz. jar instant coffee or tea
- two packages of crackers
- two pounds of sugar
- two quarts of water per person

Special precautions for apartments/condominiums

- Make one person the building captain to supervise storm preparation.
- Know your exits.
- Locate safest areas for occupants to congregate.
- Close, lock, and tape windows.
- Remove loose items from terraces (and from your absent neighbors' terraces).
- Remove or tie down loose objects from balconies or porches.
- Assume other trapped people may wish to use the building for shelter.

Special precautions for mobile homes

- Pack breakables in padded cartons and place on floor.
- Remove bulbs, lamps, mirrors—put them in the bathtub.
- Tape windows.

_ Turn off water, propane gas, electricity.
_ Disconnect sewer and water lines.
_ Remove awnings.

Special precautions for businesses

_ Take photos of building and merchandise.
_ Assemble insurance policies.
_ Move merchandise away from plate-glass windows.
_ Move merchandise to as high a location as possible.
_ Cover merchandise with tarps or plastic.
_ Remove outside display racks and loose signs.
_ Take out lower file drawers, wrap in trash bags, and store high.
_ Sandbag spaces that may leak.
_ Take special precautions with reactive or toxic chemicals.

If evacuation is advised

_ Leave as soon as you can. Follow official instructions only.
_ Follow official evacuation routes unless those in authority direct you to do otherwise.
_ Take these supplies:
  - _ change of warm, protective clothes
  - _ first-aid kit
  - _ baby formula
  - _ disposable diapers
  - _ special medicine
  - _ identification tags: include name, address, and next of kin (wear them)
  - _ flashlight
  - _ food, water, gum, candy
  - _ rope, hunting knife
  - _ waterproof bags and ties
  - _ can opener and utensils
  - _ blankets and pillows in waterproof casings
  - _ radio
  - _ fresh batteries
  - _ bottled water
  - _ purse, wallet, valuables
  - _ life jackets
  - _ games and amusements for children

_ Disconnect all electric appliances except refrigerator or freezer. Their controls should be turned to the coldest setting and the doors kept closed.
_ Leave food and water for pets. Seeing-eye dogs are the only animals allowed in shelters.
_ Shut off water at the main valve (where it enters your house).
_ Lock windows and doors.
_ Keep important papers with you:
  - _ driver's license and other identification
  - _ insurance policies
  - _ property inventory
  - _ Medic Alert or other device to convey special medical information

During the hurricane

_ Stay indoors and away from windows and glassed areas.
_ If you are advised to evacuate, *do so at once!*
_ Listen for continuing weather bulletins and official reports.

- Use your telephone only for emergencies.
- Follow official instructions only. Ignore rumors.
- Keep *open* a window or door on the side of the house opposite the storm winds.
- Beware of the "eye of the hurricane." A lull in the winds does not necessarily mean that the storm has passed. Remain indoors unless emergency repairs are necessary. Exercise caution. Winds may resume suddenly, in the opposite direction and with greater force than before. Remember, if wind direction does change, the open window or door must be changed accordingly.
- Be alert for rising water.
- If electric service is interrupted, note the time:
    - Turn off major appliances, especially air conditioners.
    - Do not disconnect refrigerators or freezers. Their controls should be turned to the coldest setting and doors closed to preserve food as long as possible.
- If you detect *gas*:
    - Do not light matches or turn on electrical equipment.
    - Keep away from fallen wires. Report location of such wires to the utility company.
    - Report gas service interruptions to the gas company.
    - Extinguish all flames.
    - Shut off gas supply at the meter.*
- Use only "safe" water:
    - The only safe water is the water you stored before it had a chance to come in contact with flood waters.
    - Should you require an additional supply, boil water for 30 minutes before use.
    - If you are unable to boil water, treat water you will need with water purification tablets.
    - An official announcement will proclaim tap water "safe." Treat all water except stored water until you hear the announcement.

After the hurricane has passed

- Listen for official word of danger having passed.
- Watch out for loose or hanging power lines and gas leaks. People have survived storms only to be electrocuted or burned.
- Walk or drive carefully through the storm-damaged area. Streets will be dangerous because of debris, undermining by washout, and weakened bridges.
- Eat nothing and drink nothing that has been touched by flood waters.

*Gas should be turned back on only by a gas serviceman or licensed plumber.

- Place spoiled food in plastic bags and tie securely.
- Dispose of all mattresses, pillows, and cushions that have been in flood waters.
- Contact relatives as soon as possible.
- If you are stranded, signal for help by waving a flashlight at night or white cloth during the day.

Appendix B
A guide to federal, state, and local agencies involved in coastal development

Numerous agencies at all levels of government are engaged in planning, regulating, or studying coastal development in Maryland and Virginia. These agencies issue permits for various phases of construction and provide information on development to the homeowner, developer, or planner. Following is an alphabetical list of topics related to coastal development. Under each topic are the names of agencies to consult for information.

Aerial photography and maps

Comparison of aerial photographs taken during different times or maps from different years is the most widely used technique to document shoreline changes. Settlements of legal disputes involving changes in shorelines frequently rely heavily on this type of documentation. Topographic maps are available from:

Distribution Section
U.S. Geological Survey
1200 South Eads Street
Arlington, VA 22202

Also available are index maps that list topographic maps. Write to the above address and ask for an index map for your area of interest. Names of local suppliers and details for ordering maps are also supplied.

The National Ocean Survey supplies nautical charts in various scales that contain navigation information and bottom depths. Index maps for these charts are available from:

National Ocean Survey
Distribution Division (C-44)
National Oceanic and Atmospheric Administration
Riverdale, MD 20840

Again, provide the geographic location of interest. The National Ocean Survey will provide the relevent information. Often both topographic and nautical charts can be bought at local nautical or outdoor supply stores. Check around.

Persons interested in aerial photography, remote-sensing imagery, or agencies that supply aerial photographs or images should contact the appropriate office listed below.

For historic listing of available photography (type, scale, year flown, coverage, percentage of cloud cover, etc.) contact:

National Cartographic Information Center
U.S. Geological Survey
507 National Center
Reston, VA 22092

Recent aerial photography should be available from:

U.S. Department of Agriculture
Agricultural Stabilization and Conservation Service
Aerial Photography Field Office
2222 West, 2300 South
P.O. Box 30010
Salt Lake City, UT 84125

Request "status of aerial photography coverage" for Maryland. Black-and-white vertical aerial photos are available for coastal counties.

For information on satellite imagery contact:

EROS Data Center
U.S. Geological Survey
Sioux Falls, SD 57198

Other, more conveniently located sources that may have aerial photographs of your area of interest available for inspection include the county divisions of the Agricultural Stabilization and Conservation Service, the office of the tax assessor in your county, departments of geology or geography in local colleges and universities, and the district office of the U.S. Army Corps of Engineers in Baltimore and Norfolk (address given in next section).

For information on flood-zone maps, see list under *Insurance* in this appendix or chapter 6.

Beach erosion and flooding in Maryland

Information on beach erosion, flooding, and general information concerning the coastal zone is available from:

Coastal Engineering Branch
Baltimore District Army Corps of Engineers
P.O. Box 1715
Baltimore, MD 21203

Coastal Resources Division
Tidewater Administration
Maryland Department of Natural Resources
Tawes State Office Building
Annapolis, MD 21401

Maryland Geological Survey
2300 St. Paul Street
Baltimore, MD 21218

Beach erosion and flooding in Virginia

Information on coastal erosion, flooding, and general information concerning the coastal zone is available from:

Coastal Engineering Branch
Norfolk District U.S. Army Corps of Engineers
803 Front Street
Norfolk, VA 23510

Virginia Marine Resources Commission
P.O. Box 756
Newport News, VA 23607

Division of Soil and Water Conservation
Shoreline Programs
P.O. Box 1024
Gloucester Point, VA 23062

Council on the Environment
903 Ninth St. Office Building
Richmond, VA 23219

Virginia Beach Erosion Council
753 General Booth Blvd.
Virginia Beach, VA 23451

Virginia Institute of Marine Science
Gloucester Point, VA 23062

Building permits

A variety of permits may be required depending upon the construction site you select. The process is further complicated by overlapping jurisdictions and special requirements when building in the coastal zone. Probably the best general advice is to contact your local building inspector's office for help in guiding you through the process of acquiring permits.

Consultants

It is inappropriate for the authors of this publication to endorse any individual or firm as a recommended coastal or construction consultant. We do, however, encourage prospective buyers and owners of existing property to seek expert advice on housing construction safety and site safety with respect to coastal hazards. The offices listed below under other topics and the offices of your local government are sources of advice on appropriate private consultants for your particular problem. In addition, state colleges and universities, particularly those with coastal geologists and coastal engineers such as the University of Maryland, Johns Hopkins University, or conservation organizations such as the Chesapeake Bay Foundation may be able either to provide assistance or names of people who can.

Dredging, filling, and construction in coastal waterways

Maryland and Virginia require that all those who wish to dredge, fill, place any structures, or otherwise alter marshlands, bay bottoms, or tidelands apply for a permit.

For information write or call:

U.S. Army Corps of Engineers
Permits Section
P.O. Box 1715
Baltimore, MD 21203

Coastal Resources Division
Tidewater Administration
Maryland Department of Natural Resources
Tawes Building, C-2
Annapolis, MD 21401

U.S. Army Corps of Engineers
Permits Section
803 Front Street
Norfolk, VA 23510

Virginia Marine Resources Commission
Habitat Management Division
P.O. Box 756
Newport News, Va 23608

Hurricane information

The National Oceanic and Atmospheric Administration is the best agency from which to request information on hurricanes. NOAA storm evacuation maps are prepared for vulnerable areas. To find out whether a map is available for your area, call or write:

Distribution Division (C-44)
National Ocean Survey
National Oceanic and Atmospheric Administration
Riverdale, MD 20840

Insurance

In coastal areas special building requirements must often be met in order to obtain flood or windstorm insurance. To find out the requirements for your area, check with your insurance agent. Further information is available from:

Federal Emergency Management Agency
Federal Insurance Administration
Washington, DC 20472

Federal Emergency Management Agency
Region III
Curtis Building
Sixth and Walnut Street
Philadelphia, PA 19106

For V-zone coverage or to request individual-structure rating contact:

National Flood Insurance Program
Attn: V-zone Underwriting Specialist
6430 Rockledge Drive
Bethesda, MD 20817

For map requests, forms, and other materials contact:

National Flood Insurance Program
P.O. Box 34294
Bethesda, MD 20817

Your insurance agent or community building inspector should be able to provide you with information about the location of your building site on the Flood Insurance Rate Map (FIRM), and the elevation required for the first floor to be above the 100-year–flood level. If they cannot provide this information, request the FIRM for your area from FEMA. Note that a flood policy under the National Flood Insurance Program is separate from your regular homeowner's policy.

Sewage and waste

For information, call or write:

State Health Department
109 Governor St.
James Madison Building
Richmond, VA 23215

Maryland Department of the Environment
201 West Preston Street
Baltimore, MD 20201

Or contact your local office.

Vegetation

Information concerning the use of marine grasses for the control of shoreline erosion can be obtained through:

Coastal Resources Division
Tidewater Administration
Maryland Department of Natural Resources
Tawes State Office Building
Annapolis, MD 21401

Marine Resources Commission
Habitat Management Division
P.O. Box 756
Newport News, VA 23608

Water resources

For information, call or write:

Water Resources Administration
Maryland Department of Natural Resources
Tawes State Office Building
Annapolis, MD 21401

State Water Control Board
2101 Hamilton Street
Richmond, VA 23230

Weather Service

General information can be obtained from:

National Weather Service
Eastern Region
585 Stewart Avenue
Garden City, NY 11530

Hurricane information is available from:

National Oceanic and Atmospheric Administration
Office of Ocean and Coastal Resources Management
3300 Whitehaven Street, N.W.
Washington, DC 20235

Appendix C
Useful references

The following publications are listed by subject. A brief description of most references is provided, and sources where the reference may be obtained are included. Some of the references are technical, but many are written in nontechnical terms for the public. Many of the publications are either low in cost, free, or readily available through libraries. We encourage readers to pursue their interests in the various topics and take advantage of these informative publications.

History and general background

1. *Tidewater Virginia*, by Paul Wilstach, 1929. This book (325 pp.) is an earlier example of a history guidebook intended for the traveler to the same historic points of interest we would visit today, but before construction of some of the great transportation links between the Eastern and Western Shores. The author notes that much of the historic soil of Jamestown lies below water, and that a seawall was built to preserve the remnant of the historic town site from further erosion. Published by the Bobbs-Merrill Co., out of print, available through libraries.

2. *Colonial Maryland*, by A. C. Land, 1981. Part of a 13-volume set entitled "A History of the American Colonies," this 367-page volume includes a brief account of the 1667 storm and describes early exploration. Published by KTO Press, Millwood, NY.

3. *Chesapeake Bay*, by the U.S. EPA Chesapeake Bay Program, 1983, a five-volume study that covers the gamut of topics from geologic history through ecology to the Bay's future. Environmental change as a result of pollution is a central theme, and the role of nutrients, toxics, and other pollutants are examined. The loss of submerged aquatic vegetation is of particular concern. The volumes include *Chesapeake Bay: Introduction to an Ecosystem*; *Chesapeake Bay Program Technical Studies: A Synthesis*; *Chesapeake Bay: A Profile of Environmental Change*; *Chesapeake Bay: A Framework for Action*; *Chesapeake Bay Program: Findings and Recommendations*. These volumes are recommended reading for planners as well as all persons generally concerned about the quality and future of the Bay. Available from the U.S. Environmental Protection Agency, Region 3, Sixth and Walnut Streets, Philadelphia, PA 19106.

4. *Chesapeake Bay Future Conditions Report*, by the U.S. Army Corps of Engineers, 1974. Of particular interest in this 12-volume

set is volume 1, *Summary*, and volume 8, *Navigation, Flood Control, and Shoreline Erosion*. The impacts of the sea-level rise, storms and storm-surge flooding, and shoreline erosion are documented and put into perspective. Although dated, this extensive report is still good reading for planners, managers, and students of the Bay. Available through larger libraries and some planning offices.

5. *Beautiful Swimmers: Watermen, Crabs and the Chesapeake Bay*, by William Warner, 1976. This Pulitzer Prize–winning book is recommended to all Bay-area residents, visitors, nature lovers, and those with an affection for seafood, particularly crabs. While the book provides a natural history of the Atlantic blue crab and captures the effort and dedication of the Bay's watermen, it provides a lesson on the Bay—its origin, function, impact on man, and man's impact on the Bay. Published by Little, Brown and Company, Boston, MA, 304 pp., and available in paperback in a 1977 Penguin edition. Available through your local bookstore.

6. *Wye Island*, by Boyd Gibbons, 1977, traces the story of conflict over the development of Wye Island on Chesapeake's Eastern Shore. It captures people's behavior and feelings during a typical battle between the big residential developers and local opponents. This 227-page nontechnical book was published with the support of Resources for the Future by Johns Hopkins University Press, Baltimore, MD 21218, and is available through bookstores and libraries.

7. *Chesapeake*, by James Michener, 1978. Sometimes fiction is as instructive as fact, and this best-selling novel lays down a story line that is based on history. Michener's characters and places are fictional, but the story of environmental decline and the disappearance of an island in the climax storm is like reality. Entertaining reading centered on Maryland's Eastern Shore. Published in hardcover by Random House and in paperback by Fawcett and available through bookstores.

8. *Chesapeake Waters: Pollution, Public Health, and Public Opinion 1607–1972*, by J. Capper, G. Power, and F. R. Shivers, Jr., 1983. Prepared mainly to provide insight into the political questions dealing with water quality, this 188-page account is filled with historical facts concerning many aspects of the Bay. Published by Tidewater Publishers, Centreville, MD 21617. Available through bookstores.

Sea-level changes

9. *Holocene Sea Level Changes and Coastal Stratigraphic Units on the Northwest Flank of the Baltimore Canyon Trough Geosyncline*, by D. F. Belknap and J. C. Kraft, 1977. Changes in sea level in the Delaware Bay region since the last major glaciation are presented in this scientific article. Published in the *Journal of Sedimentary Petrology*, vol. 47, pp. 610–29. Available through most university libraries.

10. *Sea Level Variations for the United States 1855–1980*, by S. D. Hicks, H. A. Debaugh, Jr., and L. E. Hickman, Jr., 1983. Trends in sea level since the mid-1800s to the present based on tide-gauge records are given in this government publication. Pub-

lished by the U.S. Department of Commerce, National Oceanic and Atmospheric Administration, National Ocean Service, Rockville, MD 20852.

11. *Late Quaternary Sea-Level Curve: Reinterpretation Based on Glaciotectonic Influence*, by W. P. Dillon and R. N. Oldale, 1978. An evaluation of the sea-level history of the region between Chesapeake Bay and Long Island is given in this article from a scientific journal. Published in *Geology*, vol. 6, pp. 56–60. Available through most university libraries.

Storms and storm surges

12. *Tropical Cyclones of the North Atlantic Ocean, 1871–1977*, by C. J. Neumann, G. W. Cry, E. L. Caso and B. R. Jarvinen, 1978. This report gives useful information concerning the characteristics and classification of tropical cyclones. Also included are storm tracks. Published by the National Oceanic and Atmospheric Administration, 170 pp., available through the Superintendent of Documents, Washington, DC 20402 (Stock no. 003-017-00425-2).

13. *Atlantic Hurricane Frequencies along the U.S. Coastline*, by R. H. Simpson and M. B. Lawrence, 1971. Published as a NOAA Technical Memorandum, NWS SR-58.

14. *Chesapeake Bay Tidal Flooding Study*, by the U.S. Army Corps of Engineers, Baltimore District, 1984. The report reviews the causes of coastal flooding in Chesapeake Bay and assesses flood-prone communities. It is probably available for purchase or review from the U.S. Army Corps of Engineers, Baltimore District, Baltimore, MD.

15. *Early American Hurricanes, 1492–1870*, by D. M. Ludlum, 1963. An excellent summary of the stormy history of the Atlantic and Gulf coasts that provides a lesson on the frequency, intensity, and destructive potential of hurricanes. Published by the American Meteorological Society, Boston, MA. Available in public and university libraries.

16. *The Hurricane and Its Impact*, by R. H. Simpson and H. Riehl, 1981, is an up-to-date treatment of the greatest of coastal hazards. Chapters include discussions of origin, impact of winds, waves, and tides, assessment and risk reduction, awareness and preparedness, prediction and warning, plus informative appendixes. The volume should be in libraries of coastal communities. Published by Louisiana State University Press, Baton Rouge, LA 70803.

17. *Forecasting Extratropical Storm Surges for the Northeast Coast of the United States*, by N. A. Pore, W. S. Richardson, and H. P. Perrotti, 1974. This 70-page technical report describes nor'easters and the associated storm surge in nearshore areas. Several individual storms are described, and the Ash Wednesday storm of March 1962 receives particular attention. Published by the National Weather Service as NOAA Technical Memo NWS TDL-50 and available through larger libraries.

18. *Atlantic Hurricanes*, by G. E. Dunn and B. I. Miller, 1960. This text discusses hurricanes and related phenomena such as wind, storm surge, and sea action. An appendix includes lists

of storms by region, and hurricanes of the 1950s are discussed, particularly Hurricane Hazel, 1954. Published by the Louisiana State University Press, Baton Rouge, LA 70803. Available through libraries.

19. *Chesapeake Bay Hurricane Surges*, by N. A. Pore, 1960. This older technical paper concludes that hurricanes that pass east of the Bay produce the highest surges in the southern Bay, whereas hurricanes passing west of the Bay produce the highest surges in the northern Bay. Later modeling of hurricane surges has modified these conclusions somewhat. Published in the journal *Chesapeake Science*, vol. 1, pp. 178–86. Available through college and university libraries.

20. *Hurricane Surge Predictions for Chesapeake Bay*, by C. L. Bretschneider, 1959. This older report is of interest because it includes information on several important storms including the 1933 hurricane, Hazel in 1954, Connie in 1955, and Diane in 1955. Published as U.S. Army Corps of Engineers Beach Erosion Board Miscellaneous Paper 3-59, 53 pp., available through some university libraries.

21. *Hurricanes and Coastal Storms: Awareness, Evacuation, and Mitigation*, edited by E. J. Baker, 1980, is a collection of reports addressing the problems associated with coastal storms. The paper *Living with Coastal Storms: Seeking an Accommodation*, by Richard A. Frank, pp. 4–10, is of particular interest. Available from the Marine Advisory Program, University of Florida, Gainesville, FL 32611.

22. *Storms, People and Property in Coastal North Carolina*, by Simon Baker, 1978. Although this publication is aimed at property owners on North Carolina's barrier islands, it is equally appropriate for the open-ocean coast of Virginia. Topics include storm frequency and preparing for storm impact. Sea Grant Publication UNC-SG-78-15, available from UNC Sea Grant, North Carolina State University, Raleigh, NC 27695.

23. *Hurricane Information and Atlantic Tracking Chart*, by the National Oceanic and Atmospheric Administration, 1974. A brochure that describes hurricanes, defines terms, and lists hurricane safety rules. Outlines method of tracking hurricanes and provides a tracking map. Available from the Superintendent of Documents, U.S. Government Printing Office, Washington, DC 20402.

24. *Hurricane Survival Checklist* is a free publication available from the Insurance Information Institute. Send a self-addressed, stamped business-sized envelope to Publications Service Center, Insurance Information Institute, 110 William Street, New York, NY 10038.

25. *A Storm Surge Model Study*, by J. D. Boon, C. S. Welch, H. S. Chen, R. J. Lukens, C. S. Fang, and J. M. Zeigler, 1978. This extensive report discusses causes of storm surges in Chesapeake Bay, reviews major storms, and predicts storm-surge heights along the main stem of the Bay. Although the tributaries are omitted, this report has valuable information for all areas of Chesapeake Bay. Prepared for the Federal Insurance Administration, Department of Housing and Urban Development, Washington, DC.

26. *The Strategic Role of Perigean Spring Tides in Nautical History and North American Coastal Flooding, 1635–1976*, by F. J. Wood, 1978. A 538-page report reviews tides and tidal forces with emphasis on the maximum spring tides. Spring tides, also called perigean tides, occur when the sun and moon's gravitational pull on the earth is at a maximum due to the alignment of the earth, moon, and the sun, and the moon is at its minimum distance from the earth. Although the report is largely technical, a nontechnical discussion and historical review is included. The Ash Wednesday 1962 nor'easter is one of many examples provided. Coastal-zone managers, planners, and developers should find this reference to be of particular interest because it clearly relates the likelihood and impact of storms occurring on such tides. Prepared for the National Oceanic and Atmospheric Administration, U.S. Department of Commerce and for sale from the Superintendent of Documents, U.S. Government Printing Office, Washington, DC 20402 (Stock no. 003-017-00420-1).

Geology and oceanography

27. *The Chesapeake Bay: Geology and Geography*, by M. G. Wolman, 1968. This is a part of the second volume of the report of the Governors Conference on Chesapeake Bay. Bay history and development have been influenced strongly by the natural landscape and the regional geology. The dynamic character of the Bay and its environments must be understood in light of the physical setting and history as well as the biologic setting. This technical essay is one of the best in relating the Bay region's geologic history to present processes and problems.

28. *The Virginia Coast Reserve Study: Ecosystem Description*, vol. 1, by G. J. Hennessey, 1976. The section titled "Geology" of this report (pp. 109–382) provides a review of the geology of the southern Delmarva Peninsula as well as a detailed geologic origin and description of each of Virginia's barrier islands south of Assawoman Island. Although technical in presentation, the study is a definitive review of Virginia's ocean coast and will be of interest to all students of barrier islands. A limited number of copies of the report were produced by the Nature Conservancy, 1800 N. Kent Street, Arlington, VA 22209.

29. *Geological Investigations, Chester River Study*, vol. 2, by H. D. Palmer, 1972. Published by the Maryland Department of Natural Resources.

30. *Coastal Plain Geology of Southern Maryland*, by J. D. Glaser, 1968. This somewhat-dated report and field guide still provides a good overview of coastal plain stratigraphy and includes a geologic map of southern Maryland. Published as Guidebook No. 1 of the Maryland Geological Survey, The Rotunda, 711 West 40th Street, Baltimore, MD 21211.

31. *Geography and Geology of Maryland*, by H. E. Vokes, 1956. This 243-page text provides a general treatment of geology and topography. Published as Bulletin 19 of the Maryland Geological Survey (see above).

32. *Our Changing Coastlines*, by F. P. Shepard and H. R. Wanless, 1971. This text provides an extensive review of the U.S.

coastlines, but chapter 4, pp. 87–103, is of special interest because it reviews the work on the Delmarva barrier islands and the shores of Chesapeake Bay. In particular, the rapid erosion of Tilghman Island, Poplar Island, and James Island, and the disappearance of Sharps Island are documented. Published by McGraw-Hill, New York, NY, 579 pp. Available through most college and university libraries.

33. *Sea Level Changes in the Chesapeake Bay during Historic Times*, by N. L. Fromer, 1980. This technical paper reports an average sea-level rise of 27.4 centimeters per century, or about 11 inches. Published in *Marine Geology*, vol. 36, pp. 289–305. Available through college and university libraries.

34. *Tidal Wave Characteristics of Chesapeake Bay*, by S. D. Hicks, 1964. This technical paper provides a summary of tidal conditions. Published in *Chesapeake Science*, vol. 5, no. 103–113. Available through college and university libraries.

35. *Regional Investigations of Vertical Crustal Movements in the U.S., Using Precise Relevelings and Mareograph Data*, by S. R. Holdahl and N. M. Morrison, 1974. This paper reports annual subsidence rates in the Chesapeake Bay area of -1.2 millimeters to -4.0 millimeters per year. A technical paper published in *Tectonophysics*, vol. 23, pp. 373–90. Available through college and university libraries.

Beaches and shoreline evolution

36. *The Encyclopedia of Beaches and Coastal Environments*, edited by M. L. Schwartz, 1982. This is a good source book for information and answers to coastal questions, but it is an expensive text and is aimed at the more serious student of the coast. Published by Hutchinson Ross, Stroudsburg, PA, and available through most college and university libraries.

37. *Edge of the Sea*, by Russell Sackett, 1983. This volume in the Time-Life Planet Earth Series outlines the importance and fragility of beaches and barrier systems. Coastal processes, the buffer-zone effect, the significance of coastal breeding grounds, and human impact on these environments are outlined in a nontechnical presentation. Available through your bookstore or from Time-Life Books, 541 North Fairbanks Court, Chicago, IL 60611.

38. *Waves and Beaches*, by Willard Bascom, 1980. A discussion of beaches and coastal processes. Published by Anchor Books, Doubleday, Garden City, NY 11530. Available in paperback from local bookstores.

39. *Beaches and Coasts*, by C. A. M. King, 1972 (2nd edition). Classic treatment of beach and coastal processes. Published by St. Martin's Press, 175 Fifth Avenue, New York, NY 10010.

40. *Beach Processes and Sedimentation*, by Paul Komar, 1976. Up-to-date technical explanations of beaches and beach processes. Recommended for serious students of the beach only. Published by Prentice-Hall, Englewood Cliffs, NJ 07632.

41. *The Beaches Are Moving: The Drowning of America's Shore-*

line, by Wallace Kaufman and Orrin Pilkey, Jr., 1979. This highly readable account of the state of America's coastline explains natural processes at work at the beach, provides a historical perspective of man's relation to the shore, and offers practical advice on how to live in harmony with the coastal environment. Originally published by Anchor Books/Doubleday, it is available as a 1983 paperback edition with an epilogue by the authors from Duke University Press, 6697 College Station, Durham, NC 27708.

42. *Coastal Dunes: Their Function, Delineation, and Management*, by Paul Gares, Karl Nordstrom, and Norbert Psuty, 1979. This technical report provides information about the basic factors that influence dune formation. It also provides a methodology for defining the boundaries of a management zone along the shoreline in which land use would be controlled, and it discusses the rationale for this nonengineering approach to coastal protection. Available from the Center for Coastal and Environmental Studies, Rutgers University, New Brunswick, NJ 08903.

43. *The Beachwalker's Guide: The Seashore from Maine to Florida*, by E. Ricciuti, 1982. Easy-to-read description of the origins of the Atlantic Coast's beaches, dunes, and marshes. Includes drawings, photographs, and descriptions of the dominant floral and faunal groups and species found along the Atlantic Coast. Published by Doubleday, Garden City, NY.

44. *Barrier Beach Development: A Perspective on the Problem*, by S. P. Leatherman, 1981. This article in *Shore and Beach* magazine (vol. 49, no. 2, pp. 2–9) is an excellent statement of the problems of building in the coastal zone. The brief, nontechnical discussion includes an outline of the federal government's role and offers recommendations.

45. *A Regional Test of the Bruun Rule on Shoreline Erosion*, by Peter Rosen, 1978, is a technical report demonstrating that sea-level rise can account for all of the shoreline erosion in the Bay system. The actual agents of erosion are waves, tides, storm surge, and groundwater flow. Published in *Marine Geology*, vol. 26, pp. M7–M16. Available through college and university libraries.

46. *Increasing Shoreline Erosion Rates with Decreasing Tidal Range in the Virginia Chesapeake Bay*, by Peter Rosen, 1977. This short technical paper seems to contradict what one might expect at first, but in the Bay the larger tidal range produces a higher elevation beach. Such a beach provides a more effective buffer against erosion. Published in *Chesapeake Science*, vol. 18, pp. 383–86. Available through college and university libraries.

47. *A Case Study of Littoral Drift Based on Long-Term Patterns of Erosion and Deposition*, by J. R. Schubel and others, 1972. This technical paper illustrates the concept of sediment budgets or sand supply. Knowledge of the amount of sand moved by longshore currents and their direction is essential to understanding coastal erosion and possible solutions. Published in *Chesapeake Science*, vol. 13, no. 2, pp. 80–86. Available through college and university libraries.

48. *Land against the Sea*, by the U.S. Army Corps of Engineers, 1964. Readable introduction to coastal geology and shoreline processes. Available as Miscellaneous Paper No. 4-64 from the U.S. Army Corps of Engineers.

Barrier islands

49. *Barrier Islands from the Gulf of St. Lawrence to the Gulf of Mexico*, edited by Steve Leatherman, 1979. A collection of technical papers (325 pp.) summarizing some of the geological research on barrier islands, and recommended reading for students interested in barrier islands. Published by Academic Press and available through most college and university libraries.

50. *Barrier Island Genesis: Evidence from the Central Atlantic Shelf*, by Don Swift, 1975. Technical discussion of the origin of Atlantic Coast barrier islands and their migration due to a rising sea level. Published in *Sedimentary Geology*, vol. 14, pp. 1–43, a journal likely to be found only in college and university libraries.

51. *Handbook of Coastal Processes and Erosion*, edited by Paul Komar, 1983. A collection of technical papers (305 pp.) dealing with a variety of coastal processes and environments. Of particular interest are chapter 5, *Barrier Islands*, by Dag Nummedal; and chapter 6, *Patterns and Prediction of Shoreline Change*, by Robert Dolan and Bruce Hayden of the University of Virginia. The former outlines barrier-island dynamics, and the latter provides a summary of shoreline changes for Atlantic Coast barrier islands including Virginia and the Outer Banks. Published by CRC Press, Boca Raton, FL 33431.

52. *Barrier Islands and Beaches*, 1976. Proceedings of a May 1976 barrier-islands workshop. A collection of technical papers prepared by scientists studying islands. Provides a readable overview of barrier islands. Comprehensive coverage—from aesthetics to flood insurance—by the experts. Topics include island ecosystems, ecology, geology, politics, and planning. Good bibliographic source for those studying barrier islands. Available from the Publications Department, Conservation Foundation, 1717 Massachusetts Avenue, N.W., Washington, DC 10036. Also request the foundation's free list of publications.

53. *Barrier Island Handbook*, by S. P. Leatherman, 1982 (2nd edition). A well-written introduction to barrier islands, suitable for the nonscientist. Processes, environments, coastal evolution, recreational and developmental impacts are explained and well illustrated. The 109-page book is for sale by Coastal Publications, 5201 Burke Drive, Charlotte, NC 28208.

Coastal environments

54. *Coastal Sedimentary Environments*, edited by R. A. Davis, 1985. Provides a background of knowledge of coastal environments for scientists, engineers, planners, managers, and other interested persons. Oriented toward students of geology, but has a good mix of technical information and general description that make it of use to anyone with a serious interest in coastal processes. Published by Springer-Verlag, New York, NY.

55. *Life in the Chesapeake Bay*, by A. J. Lippson and R. L. Lippson, 1984, is a well-illustrated field guide to all kinds of life found in the Chesapeake, from clams to cord grass and from stripers to barnacles. This book should be useful to fishermen,

nature lovers, or anyone who spends time near the Bay's waters. More than just a field guide, the book also provides valuable insight into the ecology of the Bay. Published by the Johns Hopkins University Press, Baltimore, MD 21218. Available at local bookstores and libraries.

56. *Wetlands Guidelines*, prepared by Virginia Institute of Marine Science and the Virginia Marine Resources Commission. This 57-page booklet is designed to provide information to Virginia citizens concerning the nature and value of wetlands ecology. It includes descriptions of natural systems and effects of different engineering practices. Although factually oriented, it is clearly written and easy to understand. Available from the Environmental Affairs Division, Virginia Marine Resources Commission, Newport News, VA 23607.

57. *Recreation in the Coastal Zone*, 1975. A collection of papers presented at a symposium sponsored by the U.S. Department of the Interior, Bureau of Outdoor Recreation, southeast region. Outlines different views of recreation in the coastal zone and the approaches taken by some states to recreation-related problems. The symposium was cosponsored by the Office of Coastal Zone Management. Available from that office, National Oceanic and Atmospheric Administration, 3300 Whitehaven Street, N.W., Washington, DC 20235.

58. *Coastal Recreation: A Handbook for Planners and Managers*, by Robert Ditton and Mark Stephens, 1976. Intended to provide technical assistance to planners and managers on major issues in coastal recreation, but may be of interest to property owners to help understand the public needs that may conflict with their rights to use the land. Published by the U.S. Department of Commerce, National Oceanic and Atmospheric Administration, Office of Coastal Zone Management, 3300 Whitehaven Street, N.W., Washington, DC 20235.

59. *A Cruising Guide to the Chesapeake*, by Fessenden S. Blanchard and William T. Stone, 1973, is a guide for the boat owner who wants to explore the Bay by water. This is similar to "cruising guides" written for other bodies of water and has been updated regularly since 1950. Published by Dodd, Mead and available at local bookstores and libraries.

60. *The Audubon Society Field Guide to North American Seashells*, by Harold A. Rehder, 1981. This well-illustrated reference is an excellent handbook for the serious shell collector. Published by Alfred A. Knopf and available in most bookstores.

61. *Seashells of North America: A Guide to Field Identification*, by R. Abbott and G. Sandstrom, 1968. A brief description of the common shells of North America, including those common to the East Coast of the United States. Grouped by family for ease of identification and illustrated with color drawings. Published by Golden Press, New York, NY, 280 pp.

62. *A Field Guide to the Birds of Eastern and Central North America*, by Roger Tory Peterson, 1980 (4th edition). Provides brief species descriptions and detailed range maps for birds of the eastern United States, including shorebirds, wading birds, and waterfowl common to the middle-Atlantic region. Illustrated with color drawings. Published by Houghton Mifflin, Boston, MA.

63. *Common Plants of the Mid-Atlantic Coast*, by Gene M. Silberhorn, 1982. A useful field guide for the identification of common plants of the coastal zone along the middle Atlantic region. Published by the Johns Hopkins University Press, Baltimore, MD.

64. *Top Flight Waterfowl Field Guide*, by J. Rutheren and W. Zimmerman, 1979. Provides color illustrations of North American waterfowl species by color-coded groupings. Illustrations allow comparison of similar males and females and includes pictures of each species in flight. Published by Moebiur Publishing Company, Milwaukee, WI.

65. *Coastal Waterbird Colonies: Cape Elizabeth, Maine to Virginia*, by M. Erwin, 1979. An inventory of seabird and wading bird nesting colonies along the northeast coast of the United States. Published by the U.S. Fish and Wildlife Service Office of Biological Services.

66. *Coastal Waterbird Colonies: Maine to Virginia*, by M. Erwin and C. Korschgen, 1977. An atlas of coastal seabird and wading bird colonies along the northeast coast of the United States. Contains maps of colony locations and information on the nesting history of species at those locations. Published by the U.S. Fish and Wildlife Service, Office of Biological Services, Washington, DC.

67. *The Audubon Society Field Guide to North American Fishes, Whales, and Dolphins*, by H. Boschung, J. Williams, D. Gotshall, D. Caldwell, and M. Caldwell, 1983. Provides detailed species accounts for fish and marine mammals of North America, including those common in the middle-Atlantic region. Illustrated with color photographs. Published by Alfred A. Knopf, New York, NY.

68. *Field Guide to Saltwater Fishes of North America*, by A. McClane, 1978. Detailed family and species accounts for saltwater fishes common to North America, including those found off the middle-Atlantic coast. Illustrated with color drawings. Published by Holt, Rinehart and Winston, New York, NY.

69. *The Complete Shellfisherman's Guide: Maine to Chesapeake Bay*, by D. Tedone, 1981. Describes habitat, habits, and harvest of recreationally important shellfish species taken from Maine to the Chesapeake Bay. Includes recipes for each species, maps of general distribution, and a chapter on pollution as it concerns shellfishing. Published by Peregrine Press, Old Saybrook, CT.

70. *A Field Guide to the Atlantic Seashore from the Bay of Fundy to Cape Hatteras*, by K. Gosner, 1978. Detailed species accounts for common marine life found along the East Coast of the United States. Illustrated with color and black-and-white drawings. Published by Houghton Mifflin, Boston, MA.

71. *The Audubon Society Field Guide to North American Seashore Creatures*, by N. Meinkoth, 1981. Detailed species accounts and an overview of taxonomy of major shore animals of North America, including those common to the East Coast of the United States. Illustrated with color photographs. Published by Alfred A. Knopf, New York, NY.

72. *The Tidemarsh Guide*, by M. Roberts, 1979. A detailed description of tidal-marsh ecology and the species of plants and animals common to these habitats. Illustrated with black-and-white drawings. Published by E. P. Dutton, New York, NY.

73. *Dipping and Picking: A Guide to Recreational Crabbing*, by D. C. Smith, 1978. How to catch, clean, and cook crabs. Available from the Sea Grant Advisory Program, South Carolina Marine Resource Center, P.O. Box 12559, Charleston, SC 25412.

Shoreline engineering

74. *Shore Erosion Control: A Guide for Waterfront Property Owners in the Chesapeake Bay Area*, by U.S. Army Corps of Engineers, Baltimore District, is a 62-page report that reviews the various types of structures to combat shoreline recession. Both positive and adverse effects of structures are noted, and the various designs are reviewed.

75. *Shore Protection Manual*, by the U.S. Army Corps of Engineers, 1984 (4th edition). The "bible" of shoreline engineering outlines the various types of engineering structures, including destructive side effects. Published in three volumes and for sale from the Superintendent of Documents, U.S. Government Printing Office, Washington, DC 20402 (Stock no. 008-022-00218-9).

76. *Low Cost Shore Protection*, by the U.S. Army Corps of Engineers, 1981. A set of four reports written for the layman under this title includes the introductory report, a property-owner's guide, a guide for local government officials, and a guide for engineers and contractors. The reports are a summary of the Shoreline Erosion Control Demonstration Program and suggest a wide range of engineering devices and techniques to stabilize shorelines, including beach nourishment and vegetation. The reports are available from the Section 54 Program, U.S. Army Corps of Engineers, USACE (DAEN-CWP-F), Washington, DC 20314.

77. *Help Yourself*, by the U.S. Army Corps of Engineers, no date. Brochure addressing the erosion problems in the Great Lakes region. May be of interest to shore residents as it outlines shoreline processes and illustrates a variety of shoreline-engineering devices used to combat erosion. Free from the U.S. Army Corps of Engineers, North Central Division, 219 South Dearborn Street, Chicago, IL 60604.

78. *Bibliography of Publications Prior to July 1983 of the Coastal Engineering Research Center and the Beach Erosion Board*, by A. Szuwalski and S. Wagner, 1984. A list of published coastal research by the U.S. Army Corps of Engineers. Available free from the Coastal Engineering Research Center, U.S. Army Engineer Waterways Experiment Station, P.O. Box 631, Vicksburg, MS 39180. *List of Publications of the U.S. Army Engineer Waterways Experiment Station*, vols. 1 and 2, by R. M. Peck, 1984 and 1985, update the above reference and list publications by other research branches of the Waterways Experiment Station. Available free from the Special Projects Branch, Technical Information Center, U.S. Army Engineer Waterways Experiment Station, P.O. Box 631, Vicksburg, MS 39180.

79. *Are You Planning Work in a Waterway or Wetland?*, by the U.S. Army Corps of Engineers, Baltimore District, 1981. This 14-page pamphlet provides general information concerning federal

regulations affecting shoreline development. It is available from the U.S. Army Corps of Engineers, Baltimore District, P.O. Box 1715, Baltimore, MD 21203.

80. *Maryland Dredge and Fill Permit Process Handbook*, by the Coastal Resources Division, Tidewater Administration, Maryland Department of Natural Resources, 1983. This handbook is for those directly involved in dredge-and-fill operations in Maryland. It provides a summary of Maryland's permit process, the agencies involved at the federal, state, and local levels, and the roles played by the agencies. Available from the Department of Natural Resources and at certain libraries in the state of Maryland.

81. *U.S. Army Corps of Engineers Permit Program—A Guide to Applicants*, by the U.S. Army Corps of Engineers, 1977. This 20-page pamphlet gives relatively detailed information concerning Corps of Engineers permits, including a description of the application and review processes. Some of these procedures have been modified in order to coordinate state and federal permit processes. It is available from the nearest district office.

82. *Wetland Licenses, Permits and Notifications*, prepared for the State of Maryland Water Resources Administration, (current). This brief review of the Maryland Wetlands Act provides directions for filling out the form that serves as notification or as application to do any dredging, filling, or otherwise altering wetlands in Maryland. Available by writing to State of Maryland Water Resources Administration, Tawes State Office Building, Annapolis, MD 21401. It is a typed circular that is updated regularly. Also ask for information concerning other types of work, such as storm drain projects, if they might affect wetlands areas.

83. *State Assistance in Shore Erosion Control*, prepared for the State of Maryland Department of Natural Resources, (current). This brief information sheet outlines the state assistance programs available for property owners involved in shore-erosion–control projects. It and other information pamphlets covering shoreline-related topics can be obtained by writing to Coastal Resources Division, Tidewater Administration, Maryland Department of Natural Resources, Tawes State Office Building, C-2, Annapolis, MD 21401.

84. *Marine Contractors and Material Suppliers Directory*, prepared for the State of Maryland Department of Natural Resources, (current). This directory is a convenient listing of contractors and suppliers of materials for shoreline projects, from landscapers and pile drivers to stone products. Includes some out-of-state companies that work in the area. Does not include endorsements or advertising. Available from the Department of Natural Resources, Shore Erosion Control, Tawes State Office Building, C-3, Annapolis, MD 21401.

85. *Wetlands Protection Oriented to Chesapeake Bay—"The Virginia Picture,"* by Norman E. Larsen, 1984. This authoritative article by the assistant commissioner for environmental affairs of the Virginia Marine Resources Commission provides insight to the legal environment concerning wetlands protection in Virginia. It is about three pages long and was published in the *National Wetlands*

Newsletter, vol. 6, no. 3, 1984. This newsletter is published by the Environmental Law Institute, Suite 600, 1346 Connecticut Ave., N.W., Washington, DC 20036.

86. *Beach Nourishment along the Southeast Atlantic and Gulf Coasts*, by Todd Walton and James Purpura, 1977. Examines successes and failures of several beach-nourishment projects. In *Shore and Beach*, vol. 45, no. 3, pp. 10–18.

87. *Beach Behavior in the Vicinity of Groins*, by C. H. Everts, 1979. An interesting description of the effects of two groin fields in New Jersey that concludes that groins deflect the movement of sand seaward, causing erosion in the downdrift shadow area. This negative downdrift effect occurs even if groin compartments are filled with sand. Published in *Proceedings of the Specialty Conference on Coastal Structures*, vol. 79, pp. 853–67.

88. *Shore Protection Guidelines*, by the U.S. Army Corps of Engineers, 1971. Summary of the effects of waves, tides, and winds on beaches and engineering structures used for beach stabilization. Available free from the Department of the Army, U.S. Army Corps of Engineers, Washington, DC 20318.

89. *An Assessment of Shore Erosion in Northern Chesapeake Bay and the Performance of Erosion Control Structures*, edited by Chris Zabawa and Chris Ostrom, 1982. Different types of protective engineering structures are described and evaluated. Parts of the 295-page report are technical, but it provides good instructional reading for those considering building new or replacing old structures. Available from the Coastal Resources Division, Tidewater Administration, Maryland Department of Natural Resources, Tawes State Office Building, Annapolis, MD 21401.

Hazard evaluation

90. *Shoreline Erosion in Virginia*, by Scott Hardaway and Gary Anderson, 1980. An excellent Virginia Institute of Marine Science publication that concisely summarizes the shoreline erosion problem in Virginia and the causes. Coastal property owners in Maryland will find the publication equally applicable to their shore. Shoreline variables are reviewed and suggestions for shore protection are provided, including the use of protective vegetation. The 25-page booklet is written in nontechnical language, and is available from VIMS, Gloucester Point, VA 23062.

91. *Sedimentation and Erosion in a Chesapeake Bay Brackish Marsh System*, by J. C. Stevenson, M. S. Kearney, and E. C. Pendleton, 1985. This technical paper describes significant marsh loss at Blackwater Wildlife Refuge since the mid-1900s and documents the causes for such loss. Published in *Marine Geology*, vol. 67, pp. 213–35. Available through college and university libraries.

92. *Erosion Susceptibility of the Virginia Chesapeake Bay Shoreline*, by Peter Rosen, 1980. This technical paper describes the variables that determine how fast different parts of the Virginia shore erode. Permeable beaches, for example, provide a natural buffer against erosion. Impermeable beaches, sand over impermeable clay, have the highest average erosion rates. Marsh-coast

average erosion rates are lower. All coastal types show significant erosion rates. Published in *Marine Geology*, vol. 34, pp. 45–59. Available through college and university libraries.

93. *Coastal Mapping Handbook*, edited by M. Y. Ellis, 1978. A primer on coastal mapping outlining the various types of maps, charts, and photography available; sources for such products; data and uses; state coastal-mapping programs; informational appendixes, and examples. A valuable starting reference for anyone interested in maps or mapping, 200 pp. For sale by the Superintendent of Documents, U.S. Government Printing Office, Washington, DC 20402 (Stock no. 024-001-03046-2).

94. *Shoreline Waves: Another Energy Crisis*, by Victor Goldsmith, 1975. Shelf bathymetry is shown to be a controlling factor in wave refraction, which, in turn, controls wave-height distribution along the beach. This report suggests that wave-energy distribution may be controlled by modifying bathymetry. Free from Sea Grant College Program, Virginia Institute of Marine Science, Gloucester Point, VA 23062. Request VIMS Contribution No. 734.

95. *Flood Hazard Management Profiles*, 1984. Center for Urban and Regional Studies, University of North Carolina, Chapel Hill, NC 27514.

96. *Preparing for Hurricanes and Coastal Flooding: A Handbook for Local Officials*, by Ralph M. Field Associates, 1983. Request Publication No. 50 from the Federal Emergency Management Agency, P.O. Box 8181, Washington, DC, or order from the Superintendent of Documents, U.S. Government Printing Office, Washington, DC 20402 (Stock no. 1985-0-419-938/28).

97. *Natural Hazard Management in Coastal Areas*, by G. F. White and others, 1976. A summary of coastal hazards along the entire U.S. coast. Discusses adjustments to such hazards, and hazard-related federal policy and programs. Summarizes hazard-management and coastal-land–planning programs in each state. Appendixes include a directory of agencies, an annotated bibliography, and information on hurricanes. A useful reference recommended to developers, planners, and managers. Available from the Office of Coastal Zone Management, National Oceanic and Atmospheric Administration, 3300 Whitehaven Street, N.W., Washington, DC 20235.

98. *Building Construction on Shoreline Property*, a checklist by C. A. Collier. Homeowners and prospective buyers of coastal property will find this pamphlet to be a handy guide in evaluating location, elevation, building design and construction, utilities, and inspection. Available free from either the Marine Advisory Program, G022 McCarty Hall, University of Florida, Gainesville, FL 32611, or the Florida Department of Natural Resources, Bureau of Beaches and Shores, 202 Blount Street, Tallahassee, FL 32304.

99. *Handbook: Building in the Coastal Environment*, by R. T. Segrest and Associates, 1975. A well-illustrated, clearly and simply written book on Georgia coastal-zone planning, construction, and selling problems. Topics include vegetation, soil, drainage, setback requirements, access, climate, and building orientation. Includes a list of addresses for agencies and other sources of information. Much of the information applies to the entire Atlantic Coast. Available from the Graphics Department, Coastal Area

Planning and Development Commission, P.O. Box 1316, Brunswick, GA 31520.

100. *Floodplain Management: Ways of Estimating Wave Heights in Coastal High Hazard Areas in the Atlantic and Gulf Coast Regions*, by the Federal Emergency Management Agency, 1981. This publication is of interest to planners and is available from FEMA, Washington, DC 20472.

101. *Guidelines for Identifying Coastal High Hazard Zones*, by the U.S. Army Corps of Engineers, 1975. Report outlining such zones with emphasis on "coastal special flood-hazard areas" (coastal floodplains subject to inundation by hurricane surge with a 1 percent chance of occurring in any given year). Provides technical guidelines for conducting uniform flood-insurance studies and outlines methods of obtaining 100-year–storm-surge elevations. Recommended to planners. Available from the Galveston District, U.S. Army Corps of Engineers, Galveston, TX 77553.

Specific shoreline analyses

102. *Shoreline Situation Report*, prepared under the supervision of R. J. Byrne, J. M. Zeigler, and C. H. Hobbs, III, 1974 to 1978. This outstanding set of county atlases and summaries represent the culmination of a major study at the Virginia Institute of Marine Science. The volumes represent a primary source of detailed coastal information that every coastal property owner, potential owner, planner, developer, and builder should use. Each volume reviews erosion rates, land uses, ownership, shoreline types, and similar useful information. The set includes: Accomack County, Virginia (1975), 190 pp.; Cities of Chesapeake, Norfolk and Portsmouth (1976), 87 pp.; City of Hampton, Virginia (1975), 63 pp.; City of Virginia Beach (1978), 91 pp.; Gloucester County, Virginia (1976), 71 pp.; Lancaster County (1978), 75 pp.; Mathews County, Virginia (1975), 99 pp.; Middlesex County, Virginia (1975), 65 pp.; New Kent, King William, and King and Queen Counties (1975), 89 pp.; Northampton County, Virginia (1974), 153 pp.; Northumberland County, Virginia (1978), 86 pp.; York County, Virginia (1975), 62 pp. These reports are available for public use at the local county planning offices.

103. *Historical Shorelines and Erosion Rates*, by the Maryland Geological Survey, 1975. A comprehensive review of shoreline changes and erosion rates in Maryland's Chesapeake Bay. The set of four volumes includes Upper Eastern Shore, Caroline, Cecil, Kent, Queen Annes, and Talbot counties; Lower Eastern Shore, Dorchester, Somerset, Wicomico, and Worcester counties; Upper Western Shore, Anne Arundel, Baltimore, and Harford counties; Lower Western Shore, Calvert, Charles, Prince Georges, and St. Mary's counties. Prepared by the Maryland Geological Survey, The Rotunda, 711 West 40th Street, Baltimore, MD 21211.

104. *Feasibility Report on Flood Control, Shore Erosion Control and Navigation, Smith Island, Maryland & Virginia*, by the U.S. Army Corps of Engineers, 1980. Smith Island experiences high erosion rates, up to 12 feet per year, and has lost 1,200 acres in the 93 years prior to 1980. The August 1933 storm flooded houses at Tylerton, Ewell, and Rhodes Point. This 282-page engineering

study provides background concerning the problems and hazards at Smith Island. Available for inspection at the Baltimore District, U.S. Army Corps of Engineers, Baltimore, MD.

105. *Erosion and Sedimentation in Chesapeake Bay Around the Mouth of Choptank River*, by J. F. Hunter, 1914. As one of the earliest shoreline erosion studies in the Bay, this paper documents the rapid shoreline retreat that occurred between 1848 and 1910. Reference to older but less-accurate maps suggests that erosion in the 20th century is nothing new. Published in *Shorter Contributions to General Geology*, U.S. Geological Survey Professional Paper No. 90 (1915), pp. 7–15. Available through college and university libraries.

106. *Shore Erosion in Tidewater Maryland*, by J. T. Singewald, Jr., and T. Slaughter, 1949. The "erosion problem" is treated in this early state publication. Published as Bulletin No. 6, Maryland Department of Geology, Mines, and Natural Resources. Available through libraries.

107. *The Effects of Tropical Storm Agnes on the Chesapeake Bay Estuarine System*, by Jackson Davis, 1976. The impact of few storms has been studied in as much detail as this report provides. The various articles are technical, but students of the Bay will find a wide variety of information in this thick 639-page volume. Hydrology, geology, water quality, biology, economics, and public health are discussed. The detrimental impact to shellfish as the result of the large input of sediment and fresh water is documented. Of special interest is the article "Agnes in Maryland: Shoreline Recession and Landslides," by B. G. McMullan (pp. 216–22),

which outlines the effect of torrential rainfall on cliff coasts. Publication No. 54, Chesapeake Research Consortium. Published by Johns Hopkins University Press, Baltimore, MD 21218.

108. *A Sediment Budget for the Choptank River Estuary in Maryland, U.S.A.*, by L. A. Yarbro and others, 1983. Shore erosion is the major source of sediment input into this estuary as documented by this technical paper. Published in *Estuarine, Coastal and Shelf Science*, vol. 17, pp. 555–70. Available through college and university libraries.

109. *The Role of Boat Wakes in Shore Erosion in Anne Arundel County, Maryland*, edited by Chris Zabawa and Chris Ostrom, 1980. Although the focus of this 230-page study is on the erosional effect of wakes from smaller boats, the report provides a good review of agents of coastal erosion, controls on wave formation, the sea-level rise, storm effects, and the protective role of the beach. Available from the Coastal Resources Division, Tidewater Administration, Maryland Department of Natural Resources, Tawes State Office Building, Annapolis, MD 21401.

110. *Hollands Island*, by M. A. Dunkle, 1985, is a limited edition, 55-page narrative based on an earlier history, *Vanishing Island*, by Capt. Irving M. Parks, Sr., of Dorchester County, Maryland. The island was 5 miles long and 1.5 miles wide in 1914 with a thriving community of over 60 homes plus stores, church, school, and townhall. As the island was lost to erosion, the buildings were moved to the mainland and the community made the wise choice to retreat in the face of the rising water level. Available through the Dorchester County, MD, library.

111. *Statistical Modeling of Historic Shore Erosion Rates on the Chesapeake Bay in Maryland*, by R. K. Spoeri, C. F. Zabawa, and Bruce Coulombe, 1985. This technical paper provides an example of how quantitative modeling may improve scientists' ability to predict more specific erosion rates in the future. Published in *Environmental Geology and Water Science*, vol. 7, no. 3, pp. 171–87. Available through university and college libraries.

Vegetation

112. *Shore Stabilization with Salt Marsh Vegetation*, by P. L. Knutson and W. W. Woodhouse, Jr., 1983, and *Shore Erosion Control with Salt Marsh Vegetation*, by P. L. Knutson and M. P. Inskeep, 1982, summarize the use of coastal marsh vegetation as an erosion-control measure. Artificial plantings are often a good alternative to building protective structures against erosion of low-energy or sheltered shorelines, such as in bays, sounds, lagoons, and estuaries. These publications outline criteria for determining site suitability for planting, selection of plant types, planting procedures, and estimating costs. Available from the U.S. Army Corps of Engineers, and as an update of earlier work by W. W. Woodhouse, Jr., and others (e.g., *Propagation and Use of* Spartina alterniflora *for Shoreline Erosion Abatement*, 1976, CERC Technical Report 76-2). These reports and additional information on coastal stabilization are available from the U.S. Army Corps of Engineers, P.O. Box 631, Vicksburg, MS 39180, or write NTIS, Attn: Operations Division, 5285 Port Royal Road, Springfield, VA 22161, and request publication by title, author, and year. There is a charge for these publications.

113. *Vegetation Establishment and Shoreline Stabilization: Galveston Bay, Texas*, by J. W. Webb and J. D. Dodd, 1976. This study evaluates plants as shoreline stabilizers in a low-energy estuarine environment in Texas. The results may be pertinent to similar environments along the Chesapeake Bay shore. Available as CERC Technical Paper 76-13 from the same addresses as above.

114. *"Cape" American Beachgrass—Conservation Plant for Mid-Atlantic Sand Dunes*, 1977. Brochure on the use of beach grass to stabilize dunes. Available from the Superintendent of Documents, U.S. Government Printing Office, Washington, DC 20402.

115. *Building and Stabilizing Coastal Dunes with Vegetation* (UNC-SG-82-05) and *Planting Marsh Grasses for Erosion Control* (UNC-SG-81-09), by S. W. Broome, W. W. Woodhouse, Jr., and E. D. Seneca, 1982. These publications on using vegetation as stabilizers are available from University of North Carolina Sea Grant, P.O. Box 8605, North Carolina State University, Raleigh, NC 27695. State publication number with your request.

116. *The Dune Book: How to Plant Grasses for Dune Stabilization*, by Johanna Seltz, 1976. Brochure outlining the importance of sand dunes and means of stabilizing them through grass plantings. Available from UNC Sea Grant (address under previous reference).

117. *Dune Building and Stabilization with Vegetation*, by W. W. Woodhouse, Jr., 1978. This report includes a section on the plants and planting methods needed to build and stabilize dunes. Avail-

able from the Superintendent of Documents, U.S. Government Printing Office, Washington, DC 20402 (Stock no. 008-022-00124-7).

118. *Designing for Bank Erosion Control with Vegetation*, by P. L. Knutson, 1978. Available from U.S. Army Corps of Engineers, P.O. Box 631, Vicksburg, MS 39180.

119. *Planting Guidelines for Marsh Development and Bank Stabilization*, by P. L. Knutson, 1977. Available from U.S. Army Coastal Engineering Research Center, P.O. Box 631, Vicksburg, MS 39180.

120. *Artificial Seaweed for Shoreline Erosion Control?*, by Spencer Rogers, Jr., 1986. Available from UNC Sea Grant College Program, North Carolina State University, Raleigh, NC (UNC Sea Grant Publication UNC-SG-WP-86-4). *Report on Generalized Monitoring of Seascape® Installation at Cape Hatteras Light House, North Carolina*, by U.S. Army Corps of Engineers, Wilmington District, Wilmington, NC, 1984. A thorough review of the history and results of using artificial seaweed to mitigate coastal erosion. These reports are a must to read if this approach is being considered.

Conservation, planning, and regulation

121. *A Citizen's Guide: Maryland's Coastal Zone Management Program*, by Gary Jacobik. A 39-page booklet explaining the need for a management program and how the existing program is structured. A list of useful addresses is included. Available from the Coastal Resources Division, Tidewater Administration, Maryland Department of Natural Resources, Tawes State Office Building, C-2, Annapolis, MD 21401.

122. *Ten Critical Questions for Chesapeake Bay in Research and Related Matters*, edited by L. E. Cronin, 1983. Environmental focuses include how science information can be conveyed to the public, what guidelines are needed for dredging and placement of dredged materials, and how pollution can be dealt with. The 163-page report should be of interest to planners and those who bridge the gap between investigators and the Bay-region public. Available as Publication No. 113, Chesapeake Research Consortium, Shady Side, MD 20764.

123. *Chesapeake Bay in Legal Perspective*, by Garrett Power, 1970. Although the bulk of this report is on the legal framework of Bay management, including case studies and draft legislation, a good summary of the natural and human history is provided in an easy-to-read format. This 270-page publication is part of the Estuarine Pollution Study Series produced by the Federal Water Pollution Control Administration, U.S. Department of the Interior.

124. *The Law of the Coast in a Clamshell*, part 14: *The Maryland Approach*, by Peter Graber, 1984. This article is one of a series presenting contemporary coastal law for non-attorneys. Topics include title to lands (upland, tide land, submerged land), tidal boundaries, public-trust doctrine, access rights, private littoral rights, and leasing of coastal zone lands and waters. Published in *Shore and Beach*, vol. 52, pp. 3–10. Available through college and university libraries. An initial article presents an overview

(*Shore and Beach*, 1980, vol. 48, pp. 14–20), followed by "The Federal Government's Expanding Role" (vol. 49, pp. 16–20), and successive articles for each coastal state.

125. *Subtitle 15 Chesapeake Bay Critical Area Commission Criteria for Local Critical Area Program Development: Final Regulations*, under *Title 1: Independent Agencies, Authority: Natural Resources Article 8-1808(d), Annotated Code of Maryland*, and *A Guide to the Chesapeake Bay Critical Area Criteria*, by the Chesapeake Bay Critical Area Commission, 1986. These documents define critical areas and the regulations for managing such areas. For information on the regulations, contact the Chesapeake Bay Critical Area Commission, Tawes State Office Building, D-4, Annapolis, MD 21401.

126. *Barrier Islands*, by H. C. Miller, 1981, published in *Environment* (vol. 23, pp. 6–11, 36–42), is an excellent review of how the federal government has stimulated barrier-island development, and the resulting financial losses in tax dollars. The conclusions are equally applicable to other coastal areas, including the Bay shore.

127. *Shoreline Erosion in the Commonwealth of Virginia: Problems, Practices, and Possibilities*, by R. J. Byrne and others, 1979. This unpublished report was prepared for the Virginia Office of the Secretary of Commerce and Resources. The report examines management practices and options to mitigate the effects of erosion. Laws and policies are reviewed, and an introductory review of coastal processes, erosion problems, and how to cope with them is provided.

128. *Questions and Answers on the National Flood Insurance Program*, by FEMA, 1983. Pamphlet explaining basics of flood insurance and providing addresses of Federal Emergency Management Agency offices. Free from FEMA, Federal Insurance Administration, Washington, DC 20472.

129. *Development of the Coast: Facing the Tough Issues*, a Coastal States Organization conference held in Charleston, SC, 1979. These published final proceedings of the conference give an abbreviated overview of the wide range of problems generated by coastal development. Available from CSO, Conference Management Associates, Ltd., 1044 National Press Bldg., Washington, DC 20045.

130. *Coastal Ecosystem Management*, by John Clark, 1977. This 928-page text covers most aspects of the coastal zone from descriptions of processes and environments to legal controls and outlines for management programs. Essential reading for planners and beach community managers. Published by John Wiley and Sons, New York, NY (1983 reprint with corrections is available from Krieger Publishing Co., P.O. Box 9542, Melbourne, FL 32902).

131. *Coastal Environmental Management*, prepared by the Conservation Foundation, 1980. Guidelines for conservation of resources and protection against storm hazards including ecological descriptions and management suggestions for coastal uplands, flood plains, wetlands, banks and bluffs, dune lands, and beaches. Part 2 presents a complete list of federal agencies and their authority under the law to regulate coastal zone activities. A good

reference for planners and persons interested in good land management. Available from the Superintendent of Documents, U.S. Government Printing Office, Washington, DC 20402.

132. *Coastal Affair*, edited by Jennifer Miller, 1982. A special subject issue of *Southern Exposure*, vol. 10, no. 3, that explores a wide range of coastal issues from the physical to the social and economic. Available from *Southern Exposure*, P.O. Box 531, Durham, NC 27702.

133. *Patterns and Trends of Land Use and Land Cover on Atlantic and Gulf Coast Barrier Islands*, by H. F. Lins, Jr., 1980. U.S. Geological Survey, Professional Paper 1156. Available from the Superintendent of Documents, U.S. Government Printing Office, Washington, DC 20402, or through your local college or university library.

134. *Design with Nature*, by Ian McHarg, 1969. A classic text on the environment. Stresses that when man interacts with nature, he must recognize its processes and governing laws, and realize that it both presents opportunities and requires limitations. Published by Doubleday and Company, Inc., Garden City, NY 11530.

135. *Who's Minding the Shore?*, by the Natural Resources Defense Council, Inc., 1976. A guide to public participation in coastal-zone management. Defines coastal ecosystems and outlines the Coastal Zone Management Act, coastal-development issues, and means of citizen participation in coastal-zone management. Lists sources of additional information. Available from the Office of Coastal Zone Management, National Oceanic and Atmospheric Administration, 3300 Whitehaven Street, N.W., Washington, DC 20235.

136. *Ecological Determinants of Coastal Area Management*, by Francis Parker, David Brower, and others, 1976. Volume 1 defines the barrier-island and related lagoon-estuary systems and the natural processes that operate within them. Outlines man's disturbing influences on coastal environments and suggests management tools and techniques. Volume 2 is a set of appendixes that includes information on coastal-ecological systems, man's impact on shorelines, and tools and techniques for coastal-area management. Also contains a good bibliography. For sale from the Center for Urban and Regional Studies, University of North Carolina, 108 Battle Lane, Chapel Hill, NC 27514.

137. *The Fiscal Impact of Residential and Commercial Development, A Case Study*, by T. Muller and G. Dawson, 1972. A classic study that demonstrates that development may ultimately increase, rather than decrease community taxes. For sale from the Publications Office, the Urban Institute, 2100 M Street, N.W., Washington, DC 20037. Refer to URI-22000 when ordering.

Building or improving a home

Current and prospective owners and builders of homes in storm-prone areas should supplement the information and advice provided in this book with that offered in references dealing specifically with safe construction. These excellent references contain sound, useful information to help residents minimize the losses caused by extreme wind or rising water. Many of these publications are free. The government publications are paid for by your taxes, so why not use them? The following references are recommended to those readers who wish to investigate further the subject of storm- and flood-resistant construction.

138. *Coastal Design: A Guide for Planners, Developers, and Homeowners*, by Orrin H. Pilkey, Jr., Orrin H. Pilkey, Sr., Walter D. Pilkey, and William J. Neal, 1983. A detailed companion volume and construction guide expanding on the information outlined in this text. Chapters include discussions of shoreline types, individual-residence construction, making older structures storm worthy, high-rise buildings, mobile homes, coastal regulations, and the future of the coastal zone. Published by Van Nostrand Reinhold, New York, NY.

139. *Coastal Construction Manual*, prepared by Dames and Moore, and Bliss and Nyuitray, Inc., for the Federal Emergency Management Agency, 1986. A guide to the coastal environment with recommendations on site and structure design relative to the National Flood Insurance Program. The report includes design considerations; examples; construction costs; appendixes on design tables, bracing, design worksheets, and wood preservatives; and a list of useful references. The manual is available from the Superintendent of Documents, U.S. Government Printing Office, Washington, DC 20402 (Stock no. 620-214/40619) or contact a FEMA office and request FEMA Publication No. 55.

140. *Elevated Residential Structures*, prepared by the American Institute of Architects Foundation (1735 New York Ave., N.W., Washington, DC 20006) for the Federal Emergency Management Agency, 1984. This excellent publication outlines coastal and riverine flood hazards and the necessity for proper planning and construction. The 137-page book discusses the National Flood Insurance Program, site analysis and design, design examples, and construction techniques. It includes illustrations, glossary, references, and worksheets for estimating building costs. Available from the Superintendent of Documents, U.S. Government Printing Office, Washington, DC 20402 (Stock no. 1984-0-438-116), or contact FEMA Region III Office, Curtis Building, Sixth and Walnut Street, Philadelphia, PA 19106, and request FEMA Publication No. 54.

141. *Design Guidelines for Flood Damage Reduction*, prepared by the American Institute of Architects Foundation for the Federal Emergency Management Agency, 1981. This is the companion volume to the previous two references and is available from the same sources.

142. *Flood Emergency and Residential Repair Handbook*, prepared by the National Association of Homebuilders Research Advisory Board of the National Academy of Science, 1986. Step-by-step guide to flood-proofing and clean-up procedures and repairs including those to household goods and appliances. Available from the Superintendent of Documents, U.S. Government Printing Office, Washington, DC 20402, or from FEMA, P.O. Box 8181, Washington, DC 20024 (FEMA Publication No. FIA-13).

143. *Floodproofing Non-Residential Structures*, prepared by Booker Associates, Inc., 1986. A 199-page illustrated guide to

flood proofing buildings as a method of reducing losses due to floods. Recommended reading for planners, community officials, managers, and owners of all types of buildings located in flood zones, both coastal and riverine. Available from the U.S. Superintendent of Documents, U.S. Government Printing Office, Washington, DC 20402 (Stock no. 1986-621-393/00128), or from FEMA, P.O. Box 8181, Washington, DC 20024 (FEMA Publication No. 102).

144. *FEMA Publications Catalog*, by the Federal Emergency Management Agency, lists over 300 titles of publications available from the agency to assist everyone from individual property owners to emergency managers. Request the catalog from FEMA, P.O. Box 8181, Washington, DC 20024.

145. *A Coastal Homeowner's Guide to Floodproofing*, by the Massachusetts Disaster Recovery Team, is a booklet that offers a checklist to take the homeowner through the process of flood proofing an existing house and tips on dealing with engineers and contractors. Although prepared for New England, the guide is appropriate to some Bay construction. Available from the Office of the Lieutenant Governor, State House, Boston, MA 02133.

146. *Wind Resistant Design Concepts for Residences*, by Delbart B. Ward. With vivid sketches and illustrations, displays construction problems and methods of tying structures to the ground. Considerable text and excellent illustrations devoted to methods of strengthening residences. Offers recommendations for relatively inexpensive modifications that will increase the safety of residences subject to severe winds. Chapter 8, "How to Calculate Wind Forces and Design Wind-Resistant Structures," should be of particular interest to the designer. Available as TR-83 from the Defense Civil Preparedness Agency, Department of Defense, The Pentagon, Washington, DC 20301, or from the Defense Civil Preparedness Agency, 2800 Eastern Boulevard, Baltimore, MD 21220.

147. *Interim Guidelines for Building Occupant Protection from Tornadoes and Extreme Winds*, TR-83/A, and *Tornado Protection—Selecting and Designing Safe Areas in Buildings*, TR-83B. These are supplement publications to the above reference and are available from the same address.

148. *Standard Building Code*. Available from Southern Building Code Congress, 1116 Brown Marx Building, Birmingham, AL 35203; or Southern Building Code Publishing Company, 3617 8th Avenue South, Birmingham, AL 35222.

149. *The Uniform Building Code*. Available from International Conference of Building Officials, 5360 South Workman Mill Road, Whittier, CA 90601.

150. *The BOCA Building Code*. Available from Building Officials and Code Administrators International, Inc., 17926 South Halsted St., Homewood, IL 60430.

151. *Residential Erosion and Sediment Control*, published jointly by the Urban Land Institute, the American Society of Civil Engineers, and the National Association of Home Builders. This is a well-illustrated 60-page guide to the objectives, principles, and design considerations for reducing erosion and sediment problems. Legal implications are also covered. This publication should be

of interest to owners, builders, and planners, and is available from NAHB, 15th and M Streets, N.W., Washington, DC 20005.

152. *Coastal Construction Building Code Guidelines*, edited by R. R. Clark, 1980. Although developed for Florida, these guidelines are applicable to other coastal areas and make specific recommendations to strengthen the standard building code in coastal areas. Available as Technical Report No. 80-1 from Bureau of Beaches and Shores, Florida Department of Natural Resources, 3900 Commonwealth Blvd., Tallahassee, FL 32303.

153. *Manufactured Home Installation in Flood Hazard Areas*, prepared by the National Conference of States on Building Codes and Standards, Inc., 1985, is a 110-page guide to design, installation, and general characteristics of manufactured homes with respect to coastal and flood hazards. Anyone contemplating buying such a structure, or those already living in one, should read this publication and follow its suggestions to lessen potential losses from flood, wind, and fire. Available from the U.S. Superintendent of Documents, U.S. Government Printing Office, Washington, DC 20402 (Stock no. 1985-529-684/31054), or from FEMA, P.O. Box 8181, Washington, DC 20024 (FEMA Publication No. 85).

154. *Protecting Mobile Homes from High Winds*, TR-75, prepared by the Defense Civil Preparedness Agency, 1974. This out-of-print booklet outlines methods of tying down mobile homes and means of protection such as positioning and wind breaks.

155. *Structural Failures: Modes, Causes, Responsibilities*, 1973. See especially the chapter entitled "Failure of Structures Due to Extreme Winds," pp. 49–77. For sale from the Research Council on Performance of Structures, American Society of Civil Engineers, 345 East 47th Street, New York, NY 10017.

156. *Hurricane-Resistant Construction for Homes*, by T. L. Walton, Jr., 1976. An excellent booklet produced for residents of Florida, but equally useful to those of the Virginia and Maryland shores. A good summary of hurricanes, storm surge, damage assessment, and guidelines for hurricane-resistant construction. Technical concepts are presented on probability and its implications on home design in hazard areas. A brief summary of federal and local guidelines is given. Available from Florida Sea Grant Publications, Florida Cooperative Extension Service, Marine Advisory Program, Coastal Engineering Laboratory, University of Florida, Gainesville, FL 32611.

157. *Houses Can Resist Hurricanes*, by the U.S. Forest Service, 1965. An excellent paper with numerous details on construction in general. Pole-house construction is treated in particular detail (pp. 28–45). Available as Research Paper FPL 33 from Forest Products Laboratory, Forest Service, U.S. Department of Agriculture, P.O. Box 5130, Madison, WI 53705.

158. *Pole House Construction and Pole Building Design*. Available from the American Wood Preservers Institute, 1651 Old Meadows Road, McLean, VA 22101.

159. *Standard Details for One-Story Concrete Block Residences*, by the Masonry Institute of America. Contains nine fold-out drawings that illustrate the details of constructing a concrete-block house. Presents principles of reinforcement and good connections. This publication is aimed at designing for seismic zones, but is

applicable to hurricane zones as well. Written for both layman and designer. For sale from Masonry Institute of America, 1550 Beverly Boulevard, Los Angeles, CA 90057 (Publication 701).

160. *Masonry Design Manual*, by the Masonry Institute of America. A 384-page manual covering all types of masonry including brick, concrete block, glazed structural units, stone, and veneer. Very comprehensive and well presented. Probably of more interest to the designer than to the layman. For sale from the Masonry Institute of America, 1550 Beverly Boulevard, Los Angeles, CA 90057 (Publication 601).

161. *Model Minimum Hurricane Resistant Building Standards for the Texas Gulf Coast*. Although written specifically for the Texas coast, this publication is of interest for all coastal residents and property owners. Available from the Texas Coastal and Marine Council, P.O. Box 13407, Austin, TX 78711.

162. *Construction Materials for Coastal Structures*, by Moffatt and Nichol, Engineers, 1983, is a lengthy (427 pp.) summary of the properties and uses of a wide range of materials used in coastal structures, beach-protection devices, and erosion control. This technical reference should be of particular interest to coastal engineers and construction contractors who build such structures.

Index

Abbey Point, 117, 118
Aberdeen, Md., 3, 117
Aberdeen Proving Ground, 93
Accomack County, Va., 2, 29; gabions in, 67; shoreline in, 51, 78, 84–85; site analysis of, 86
accretion, shoreline, 76
aerial photographs, 75, 198
Amburg, Va., 137
ammophila, 31
anchoring, house. *See* construction, anchoring
Annapolis, Md., 2, 123; 100-year flood level for, 97; storm surges in, 15, 16; tides in, 9, 10
Anne Arundel County, Md., 3; erosion rates in, 124, 127; shoreline in, 97; site analysis of, 123, 125, 126
Ape Hole Creek, 88
Arlington, Va., 80
Arundal on the Bay, Md., 123
Ashland Ledge, 110
Ash Wednesday storm, 17, 22, 152
Assateague Island, 51, 57, 149, 150

Assawoman Island, 158
August storm. *See* hurricane, August 23, 1933
Avalon, Va., 132
A-zones, 157

Back Creek, 89, 114, 142
Back River, 93, 119
backwash, 30
Baltimore, 3, 77; erosion rates in, 120; storm surges in, 15, 16, 17, 90; tides in, 9
Baltimore County, 3; erosion rates in, 120; site analysis of, 118–119
Baltimore Harbor, 96; erosion rates in, 124; site analysis of, 121
barrier islands, 33–39, 211; canals on, 111; migration of, 35–39, 107; selecting a site on, 74. *See also* Delmarva barrier islands; North Carolina, Outer Banks
Battle Point, Va., 81
Batts Neck, 108
Bayford, Va., 81

Bayside, Va., 85, 140, 141
beach(es): active, 39; barrier, 31; dynamic equilibrium of, 39; erosion of, 37, 199, 200; estuarine, 29, 31; nourishment, 49, 52–53, 54, 59, 68, 152; preservation, 151; replenishment, 52–53, stabilization, 42, 43–72 passim; upper, 39; vegetation, 51
Beasley Bay, 85
Beaverlett, Va., 140
Beech Ground Swamp, 98
Belvedere Heights, Md., 3
Benoni Point, 83
berm, 42
Bertha, Md., 129
Bertrand, Va., 136
Bethel Beach, 140
Betterton, Md., 16, 90, 114
Betterton Beach, 54
Beverlyville, Va., 132
Big Annemessex River, 88, 89
Big Marsh, 85
Bivalve, Md., 91, 92

Black Walnut Point, 87
Blackwater, Md., 99
Blackwater National Wildlife Refuge, 2, 28
Blakes, Va., 137
Bloodsworth Island, 2, 79, 94–95, 96
Bloxom, Va., 85
bluff(s), 25–26, 87, 90; as evidence of eroding shoreline, 42, 73, 97; slumping of, 26, 42, 97
Bluff Point, 84, 134
Bobtown, Va., 84
Bodkin Creek, 122
Bodkin Point, 122
Bogey Neck, 132
Bohannon, Va., 141
Bohemia River, 114, 116
Bolingbroke Creek, 102
Boston, Va., 84
breakwaters, 68; experimental (Potter's fence), 152; offshore, 67; sandbars as, 12; types of, 69
Bridgetown, Va., 81
Broad Creek, 102, 104

Broad Neck, 96, 122, 123
Broadwater, Va., 148
Buckroe Beach, 111, 142
Buffers Bluff, 80
building. *See* construction
building codes, 166, 175, 179, for high-wind areas, 181, 184–187
building permits, 163, 200
bulkheads, 44, 46, 61; case histories of, 64–65; compared to revetments, 63–64; design and construction of, 59–63; end-around effects, 49, 57; ineligibility for flood insurance, 158; types of, 58
Bull Minnow Point, 114
Bull Neck, 132
Burgess, Va., 132
Bush River, 117, 118
Byrds Marsh, 85
Byrdton, Va., 134

Calvert Bay, 130
Calvert Beach, Md., 2
Calvert Cliffs, 25
Calvert County, Md., 2; erosion rates in, 127–128; shoreline in, 97; site analysis of, 126, 128
Cambridge, Md., 2, 15, 101, 105; storm-surge height in, 16
Canton, Md., 121
Cape Charles, 2, 80, 149

Cape Charles City, Va., 2, 55, 80
Cape Henry, 2, 115, 145, 147, 149
Cape Lookout, 149
Cape St. Claire, 3, 122
Cashville, Va., 84
Castle Haven, 102
Cattail Island, 99
Cecil County, Md., 3, 114, 115, 116, 120
Cedarhurst, Md., 125
Cedar Island, 149, 150, 159
Cedar Point, 130
Cedar Straits, 88
Chance, Md., 16
Chapel Neck, 141
Charles Cannon Road, 88
Charles County, Md., 2
Charles Creek, 98
Charlestown, Md., 116
Chelsea Park, Md., 3
Cheriton, Va., 80
Cherry Point, 140
Cherrystone Inlet, 80
Chesaco Park, Md., 119
Chesapeake, Va., 2
Chesapeake Bay Bridges, 108, 123
Chesapeake Bay Critical Area Commission, 161, 162
Chesapeake Bay Critical Area Law, 161, 162, 163
Chesapeake Beach, Md., 2, 16, 126

Chesapeake Beach, Va., 132
Chesapeake City County, Va., 2
Chester River, 15, 106, 108, 110, 112; erosion rates for, 111; shoreline of, 87, 90
Chestertown, Md., 3, 15, 110
Chincoteague Bay, 38, 149
Chincoteague Island, 149
Choptank River, 1, 2, 5, 15, 30, 33, 50, 55, 56, 61, 63, 101, 102; erosion rates for, 100, 103
Christs Rock, 101
Church Creek, 81, 101
Church Neck, 78
Churchton, Md., 125
City Beach, 53
Clam, Va., 85
Clean Water Act of 1977, 160
Coastal Barrier Resources Act of 1982, 156, 159
Coastal Barrier Resources Systems, 159
coastal environments, 211–216
coastal erosion. *See* erosion, coastal
Coastal Primary Sand Dune Protection Act, 160
Coastal Zone Management Act (CZMA), 160
coastal zone management plans (CZMPs), 160
Cobb Island, 148
Cobbs Creek, Va., 137

Cod Creek, 103, 130, 132
Colonial Beach, Md., 2, 68
Colonial Beach, Va., 70
Columbia Beach, Md., 15
condominiums, 157, 158, 194. *See also* construction, high-rise
conservation, 221–223
construction, anchoring, 173, 175, 179; bond beam, 183; in coastal waterways, 201; design, 167; high-rise, 187–189; masonry, 173, 181, 183; modular, 189–190; moveback, 48, 49; piers, 174; pile, 173, 174, 175, 177, 178, 187, 189; pole or stilt, 173, 178; post, 173, 175, 178; reinforcement, 182; setback, 48, 49; steel-frame, 173
Cook Point, 83
Copperville, Md., 106
Cornfield Harbor, 16
Corps of Engineers. *See* U.S. Army Corps of Engineers
Corrotoman Point, 136
Corrotoman River, 136
Corsica Neck, 110, 112
Cottage Park, Va., 145
Cove Point, 16, 129
Cow Neck, 137
Cox Creek, 108
Crab Alley Bay, 108
Crab Alley Creek, 108

Crab Alley Neck, 106
Crabbe Mill, Va., 132
Crab Neck, 142
Crab Point, 98
Craddock Neck, 84
Craddockville, Va., 84
Cranes Creek, 132
Craney Island, 67
Crisfield, Md., 2, 15, 16, 79, 88
Critical Area Law. *See* Chesapeake Bay Critical Area Law
Critical Areas Criteria, 72
Croatan Beach, 154
Crockett Town, Va., 84
Cubitt Creek, 132
currents, longshore, 8, 12, 30, 40
Currituck Sound, 42
Curtis Bay, 121
Curtis Point, 3, 125

Dames Quarter, 79, 91
Dare, Va., 142
Dares Beach, Md., 128
Deale, Md., 3, 125
Deal Island, 32, 79, 91
Deep Creek, 85
Deep Point, 110
Deep Water Point, 104
Delaware Bay, 9, 34
Delaware River, 15
Delmarva barrier islands, 149; environments on, 34; migration of, 35, 37; origin of, 33; as part of Coastal Barrier Resources System, 159; salt marshes on, 42
Delmarva Peninsula, 20, 148
deltas, 38
Denton, Md., 15
Department of the Interior. *See* U.S. Department of the Interior
Devils Reach, 110
Diamond City, N. C., 149
Diggs, Va., 140
Diggs Wharf, 141
Disaster Relief Act of 1974, 156, 159
Dixie, Va., 137
Dixondale, Va., 141
Dorchester County, Md., 2; coastline in, 79; erosion rates in, 93, 100; site analysis of, 92, 98–99, 101
dredging, 200
drowned river valleys, 7
Drummond Field, 68
Drum Point, 2, 129
dune(s), 31–32, 51; protection of, 157; storm erosion of, 41, 173
dune-building, 150
Dunton Mill, Va., 134
Dutchman Point, 125

Eastern Neck Island, 3, 90, 110, 112
East Ocean View, 53
Easton, Md., 3, 15
East River, 141
Eastville, 16
Edgewood, Md., 118
Edwardsville, Va., 132
Elk Neck, 90, 114, 116
Elk River, 90, 116
Elliotts Neck, 81
Emerson Harrison Bridge, 102
end-around effects, 46, 59, 63, 68
engineering, shoreline. *See* stabilization, shoreline
Environmental Protection Agency (EPA). *See* U.S. Environmental Protection Agency
erosion, shoreline, 4, 13; around groins, 54; by storms, 16
erosional scarps. *See* bluffs
Essex, Md., 3, 119
Essex County, Va., 2
estuaries, 7. *See also* drowned river valleys
evacuation
Ewell, Md., 79, 94
Exmore, Va., 2

Fairhaven, Md., 126
Fairlee Creek, 113
Fairmount, Md., 89
Fairport, Va., 132
Fairview, Md., 121
Farm Creek, 92
Federal Aviation Administration (FAA), 22
Federal Emergency Management Agency (FEMA). *See* U.S. Federal Emergency Management Agency
Federal Insurance Administration (FIA), 166
Federal River and Harbor Act of 1899, 159
Federal Water Pollution Control Act of 1972, 160
Fenwick Island, 57, 148
Ferry Creek, 137
Ferry Point, 98
fetch, 11, 74, 87
FHA mortgages, 158
filter cloth, 59, 66, 68
Fishermans Island, 149, 159
Fishing Bay, 92, 98
Fishing Creek, 83, 101
Flag Cove, 98
Fleet Point, 132
Fleets Bay, 134
Fleets Island, 134
floods and flooding, 4, 75; effects on buildings, 169, 172; in embayments, 78; in Fort Story, Va., 154; in Havre de Grace, Md., 93; in marshes, 79; in Maryland, 199; reduction at high elevations, 87; in Sandbridge,

Va., 151; in Virginia, 200. *See also* 100-year flood; storm surges
Flood Disaster Protection Act of 1973, 156
Flood Hazard Boundary Maps (FHBMS), 75, 156
flood insurance, 158–159
Flood Insurance Rate Maps (FIRMS), 14, 75, 157
Flood Plain Management, 156
Fort Howard, Md., 121
Fort Monroe, Va., 111, 142
Fort Story, Va., 63, 67, 154
Fox Creek, 98
Fox Hill, Va., 142
Frankie Point, 122
Freeschool Marsh, 85

Gabions, 62, 67
Gales Creek, 88, 89
Galveston, Tex., 19, 192
Gaskins Point, 16
geology and oceanography, 208–210
Gibson Island, 96, 122
Ginny Beach, 132
Gloucester County, Va., 2, 54, 107, 137, 138
Gloucester Point, 16, 53, 61
Godfrey Bay, 137
Golden Beach, Md., 2
Golden Hill, Md., 99

Good Luck, Va., 134
Goodwin, Va., 142
grading. *See* terracing
Grafton, Va., 137
Grandview Nature Preserve, 31
Grasonville, Md., 106, 110
grass(es); artificial, 51; salt-marsh, 43, 50; sea, 50, 74
gravity structures, 68
Great Lakes, 51
Great Marsh, 98
Great Wicomico River, 132
Greenbury Point, 123
greenhouse effect, 8
Grinels, Va., 137
groins, 14, 49; fields, 14, 53–57; as hard solutions, 44, 53–57; and longshore transport, 12, 45; low-profile, 53; wooden, 65
groundwater, 26, 74
Grove Neck, 90, 114
Grubin Neck, 101
Guard Shore, 16
Guinea Neck, 141
Gunpowder Neck, 117
Gunpowder River, 93, 118
Gwynn Island, 29, 107, 140

Hack Creek, 132
Hacket Point, 96, 123
Hacks Neck, 84

Hampton, Va., 2, 14, 31; erosion rates in, 146; site analysis of, 142, 144
Hampton Roads, Va., 9, 111, 144
Hampton Roads Bridge, 142
Harborton, Va., 84
Harewood Park, Md., 93, 118
Harford County, Md., 3, 93, 117, 118, 120
Harper Creek, 137
Harris River, 104
Hartfield, Va., 137
Haven Beach, 140
Havre de Grace, Md., 3, 15, 16, 93, 117
hazard evaluation, 216–218
hazard mitigation, 156
Hell Hook Marsh, 98
Hermansville, Md., 130
Highlandtown, Md., 121
high-water state, 164
Hills Bay, 137
Hills Point, 83
Hog Island, 148, 149
Hog Point, 130
Holland Point, 126, 137
Honga River, 98
Hooper Islands, 98; erosion on, 43, 79, 100; storm damage to, 83, 100; storm surges on, 16
Hooper Point, 101
Hooperville, 43, 98

Hope Point, 98
Hopewell, Md., 88
Howell Point, 114
Hudgins, Va., 140
Hughlett Point, 134
Hull Neck, 130, 132
Hungars Creek, 81
hurricanes, 18, 20; Agnes, 20; of August 23, 1933, 57, 79, 83, 90, 148, 152; Barbara, 20; Betsy, 192; Camille, 19, 172, 181, 186, 192; Charley, 151–152; Connie, 17, 20, 21; damage from, 68, 175; Diane, 17; Donna, 20, 21; Flossy, 20, 21; Frederic, 186; Gloria, 20, 21, 61; Hazel, 17, 20, 79; information, 201; precautions, 193–197; ranking, 192–193
Hyslop Marsh, 84

Ice Age, 6
Indian Creek, 107, 134
Indian Point, 104
Ingram Bay, 132
inlet(s), 38, 74, 169; shifting of, 157
insurance, 201. *See also* flood insurance
Intensely Developed Areas (IDAS), 161
Inter-Agency Agreement for Nonstructural Flood Damage Reduction, 156

Intracoastal Waterway, 150
Island Creek, 102
island migration, 37, 149, 150
Isle of Wight County, Va., 2

James City County, Va., 2, 68
James Island, 5, 83, 99, 101
James River, 2, 6, 144
James River Bridge, 67
Jamesville, Va., 81
Jeffs, Va., 142
Jenkins Creek, 102
jetties, 44, 45, 53–57, 150
Johnsontown, Va., 81
Joppatowne, Md., 3, 93, 118

Keene's Point, 98
Kent County, Md., 3, 110, 112, 114; erosion rates in, 115; shoreline in, 90
Kent Island, 87, 108, 109
Kent Point, 108
Kenwood Beach, Md., 129
Kilmarnock, Va., 134
King and Queen County, Va., 2
King Creek, 80
Kiptopeke, Va., 16, 67, 80
Kiptopeke Beach, 80
Kiptopeke Pier, 78

Laban, Va., 140
lagoon(s), 38, 111. *See also* sound(s)

Lake Okeechobee, Fla., 19
Lakes Cove, 98
Lake Shore, Md., 3
Lakesville, Md., 98
Lancaster County, Va., 2, 134; development in, 103; erosion in, 43, 135, 138; site analysis of, 136
land subsidence, 8
land use regulation, 160–62
Langford Creek, 110
Langley Research Center, 142
Laws Thorofare, 91
Lecompte Bay, 101, 102
Lee Mont, Va., 85
Lego Point, 118
Lewis Creek, 130
Lexington Park, Md., 2
Limited Development Areas (LDAS), 161
Little Annemessix River, 79, 88
Little Choptank River, 83, 99, 100, 101
Little Cobb Island, 159
Little Creek, 53, 145
Locklies, Va., 137
Locust Point, 114
London Towne, Md., 125
Long Beach, Md., 2, 129
Long Island, N.Y., 151
Long Point, 88, 91

longshore currents. *See* currents, longshore
longshore transport, 12, 13; forming spits, 31; and shoreline stabilization structures, 45, 53
Love Point, 16, 108
Lowes Point, 104
lowlands, 32, 79
low-water state, 164
Lubbock, Tex., 189
Lynnhaven Inlet, 152

Magathy Bay, 38
Magothy River, 96, 122
Manokin River, 89, 91
Marine Resources Commission, 164
Marion, Md., 88
Marshalls Beach, 132
marshes, 26–29, 79, 173; as flood protection, 29; fringe, 28, 103, 111, 115; margins, 28; salt, 38, 42, 74, 78, 103, 111; submergence of, 28; vegetation in, 27–28
marsh peat, 28
Maryland Department of Natural Resources, 63, 64, 73, 164; Tidewater Administration, 160
Maryland Geological Survey, 75
Matapeake, 16
Mathews, Va., 140

Mathews County, Va., 2; dunes in, 51; erosion rates in, 107, 138, 143; site analysis of, 137, 140
Mattapex, Md., 107
Mattawoman Creek, 81
Melton Point, 110
Messick, Va., 142
Messongo Creek, 85
Metomkin Bay, 38
Metomkin Island, 38, 149
Miami Beach, Fla., 151
Middle River, 119
Middlesex County, Va., 2, 107, 137, 138
migration, shoreline, 157. *See also* barrier islands, migration of
Miles River, 102, 104, 105, 106
Miles River Neck, 104
Miles Store, Va., 141
Milford Haven, Va., 140
Mill Creek, 132
Mill Point, 80
Mississippi, 172, 181
mobile homes, 157, 158, 168; anchoring, 186–189, 194
Mobjack Bay, 107, 111, 141
Mob Neck, 132
Monie Bay, 91
Moon, Va., 140
Mount Olive, Va., 132

Mulberry Point, 91, 104
Myer Creek, 136

Nandua Creek, 84
Nanticoke, Md., 79, 91
Nanticoke River, 79, 91, 92
Narris Bridge, 137
Narrows Ferry Bridge, 83
NASA, 149
Nassawadox Creek, 81
National Academy of Science, 8
National Flood Insurance Act of 1968, 156
National Flood Insurance Program (NFIP), 156, 163, 173, 178
National Ocean Survey, 198
National Park Service, 150
National Weather Service, 192
Nature Conservancy, 150
neap tides. See tides, astronomic
Neuman Neck, 132
Newcomb, Md., 104
New Jersey, 42, 51
New Point Comfort, 107, 140, 141
Newport News, 2, 142, 146
Nohead Bottom, 137
nor'easter(s), 16–17, 168
Norfolk, 2, 77, 142, 149; beach revitalization in, 150; erosion rates in, 146; storm surges in, 16, 17

North, Va., 141
Northampton County, Va., 2, 84; erosion in, 22, 82; shoreline in, 78; site analysis of, 80–81
North Bay, Va., 154
North Beach, Md., 2, 126
North Carolina, 48, 151, 153, 154; Outer Banks, 18, 19, 51, 150
North Carolina Sea Grant Program, 51
North East, Md., 116
northeaster(s). See nor'easter(s)
Northeast River, 90, 116
Northend Point, 111
North River, 141
Northumberland County, Va., 2; development in, 103; erosion rates in, 133, 135; site analysis of, 132, 134

Occohannock Creek, 81, 84
Occohannock Neck, 78, 81
Ocean City, Md., 149, 150; groins and jetties in, 55, 56; recreational beach in, 43
Ocean City Inlet, 57
oceanography. See geology and oceanography
Ocean Park, Va., 145
Ocean View, Va., 145, 151
Old Beach, 85

Old Plantation Creek, 80
Old Point Comfort, 111, 142
Old Town Neck, 81
Onancock, Va., 2, 84, 149
Onancock Creek, 84
100-year flood, 14–16, 21, 74, 87, 96, 156
100-year flood elevation. See 100-year flood level
100-year flood level, 75, 78, 79, 90, 93, 97, 169, 173
100-year storm, 57
Orchard Beach, Md., 3
Oregon Inlet, 42, 153
Otter Point, 117
overwash, 39; fans, 39
Owings Beach, Md., 125
Oxford, Md., 102
Oyster Cove, 99

Park Hall, Md., 130
Parramore Island, 149
Parrott Island, 137
Parson Island, 108
Parsons Creek Neck, 99
Patapsco River, 96, 121
Patuxent River, 15, 97, 129, 130
Pearce Neck, 116
perched beach, 68, 71
Perrin, Va., 141

Perryville, Md., 3, 90, 116
Persimmon Point, 88
Phoebus, Va., 142
Piankatank River, 137, 138, 140
piles. See construction, pile
Piney Neck Point, 106
Plum Point, 126, 128
Pocomoke City, Md., 15
Pocomoke Sound, 2, 85, 88, 149
Point Lookout, 2, 130
Point No Point, 97, 130
Poplar Island, 1, 5
Poquoson, Va., 142
Poquoson River, 111, 142
Port Patience, 129
Port Republic, 129
Portsmouth, 2, 67
Potomac River, 2, 15, 130; erosion along, 103; flood risk, 97; role in origin of Chesapeake Bay, 6
Poverty Point, 99
Powell's Bluff, 81
power failures, 189
Pribble, Va., 141
Prince Georges County, Md., 3
Prospect Bay, 106
Pungoteague, Va., 84

Quaker Neck, 112
Quantico Creek, 91, 92

Queen Annes County, Md., 64, 87, 106
Queensland, Australia, 168

Rappahannock River, 2, 134, 136, 137; erosion along, 43, 107, 138; role in origin of Chesapeake Bay, 6
reach(es), 45
Reed Creek, 110
Remo, Va., 132
Resource Conservation Areas (RCAS), 161, 162
revetments, 49, 57; compared to bulkheads, 64; end-around effects on, 46; filters in, 66; as hard solutions, 44; types of, 62, 87
Rhode River, 125
Rhodes Point, 79, 94
Richmond County, Va., 2
Rich Neck, 106
ridges. See sandbars
Rigby Island, 140
riparian rights, 155
riprap, 57, 61
Riverside Wharf, 110
Riviera Beach, Md., 3
Robb, Governor Charles, 153
Robins Neck, 107, 141
Robins Point, 118
Rock Hall, Md., 90, 112
Rock Hall Harbor, 112

Rocky Point, 129
rollover. See barrier islands, migration of
Romney Creek, 117
roof(s): construction, 179–181; gable roof, 181; hip roof, 181
Round Bay, 123
Ruark, Va., 137
Rudee Inlet, 152, 153, 154
runnels, 40
Russell Island, 32

Saffir-Simpson Hurricane Scale, 192, 193
St. George Island, 130
St. Jerome Creek, 130
St. Leonard Creek, 129
St. Marys City, Md., 130
St. Marys County, Md., 2, 97, 131
St. Marys River, 130
St. Michaels, Md., 3, 104, 106
Salisbury, Md., 2, 15
sandbars, 12, 14, 40, 78
Sandbridge, Va., 149, 153–154; erosion in, 42; flooding in, 151; shoreline retreat in, 43,
sand, fences, 51
Sandy Point, 93, 96, 116, 117, 122, 123
Sarah, Va., 140

Sassafras River, 90, 114
Saunders Point, 125
Savage Neck, 22–24, 78, 80
Scarborough Neck, 84
scarp(s), 73
Schley, Va., 141
Scientists Cliff, 65, 67, 97
scour. See wave(s), scour
Seaford, Va., 142
sea level, changes in, 6–10, 28, 34–37, 149, 205–206
seawalls, 43, 49, 57; ineligibility for flood insurance, 158
sedimentation control, 163
Seneca Creek, 118, 119
septic systems, 74, 153, 158
Severn, Va., 141
Severna Park, Md., 3
Severn River, 96, 97, 123, 141
sewage and waste, 202. See also septic systems
sewerage, 163
Shackleford Banks, 149
Shallow Creek, 119
Sharps Island, 1, 5
Ship Shoal Island, 149
Shooting Point, 81
shoreline analysis, 218–220
shoreline erosion. See erosion, shoreline
shoreline evolution, 209–210

shoreline stabilization structures, 45. See also breakwaters; groins; jetties
shore-parallel structures, 44, 57–69. See also breakwaters; bulkheads; revetments; seawalls
shore-perpendicular structures, 44. See also groins; jetties
sill(s), 23, 24
Silver Beach, Va., 14, 78, 81
site-analysis maps, 74, 76. See also specific site analyses in chapter 4
Slaughter Creek, 99
slumping. See bluffs
Small Business Administration loans, 158
Smith Island, 2, 149; development on, 79; environments on, 34; erosion rates on, 96; site analysis of, 94–95; storm damage to, 83
Smith Point, 132
Smithville, Md., 99
Solomons, Md., 9
Solomons Island, 16, 129
Somerset County, Md., 2, 84; coastal composition in, 79; erosion rates in, 90, 93; site analysis of, 88, 89
sound(s), 38
Sound Beach, Va., 84
Southeast Creek, 110
South Marsh Island, 2, 79, 94–95, 96
South River, 96, 123, 125

spalling, 26
Spaniard Neck, 110
Sparrow Point, 81
Sparrows Point, 121
spits, 31, 74, 78
spring tides. *See* tides, astronomic
stabilization, beach. *See* beach, stabilization
stabilization, shoreline, 32, 43–72 passim. *See also* beach, stabilization
Stafford County, 68
Stampers, Va., 137
Standard Building Code, 185
Stevensville, Md., 108
Still Pond, 113
Stingray Point, 2, 107, 137
Stony Point, 117
storm(s), 206–208. *See also* hurricane(s); nor'easter(s)
storm surge(s), 11, 12–16, 20, 30, 75, 90, 168, 173, 206–208
storm-surge level, 73, 93, 173, 187
Stove Point Neck, 137
structural engineering, 168
Suffolk City County, Va., 2
Surry County, Va., 2
Susquehanna Flats, 90
Susquehanna River, 15, 116, 117; erosion along, 93; role in formation of Chesapeake Bay, 6
Swan Point, 90, 112

swash, 30
Syringa, Va., 137

Taft, Va., 134
Talbot County, Md., 102, 103, 104, 105, 106
talus, 65, 73
Tangier Island, 2, 43, 79, 94
Tankards Beach, 78, 80
Tar Bay, 98
Taylors Island, 79, 83, 99, 100
Taylors Island, Md. (city), 99
Taylors Island Wildlife Management Area, 99
terracing, 65
The Gulf, Va., 81
tides, astronomic, 10–12; neap tides, 10; spring tides, 10, 17; wind, 11
Tidewater Administration. *See* Maryland Department of Natural Resources, Tidewater Administration
Tilghman Island, 5; erosion of, 4, 83, 87, 105; site analysis of, 104
Tilghman Point, 104
Todds Point, 102
Tolchester, 16
Tolchester Beach, 112
topographic maps, 198
Town Point Neck, 90, 114, 116
Tred Avon River, 102, 106
Trippe Bay, 83, 101

tropical storms, 16, 18–20. *See also* hurricanes
Tulls Point, 88
Turkey Point, 114, 116, 125
Twin Pines, Va., 144
Tylerton, 79, 94
Typhoon Tracy, 168, 181

undertow, 40
U.S. Army Corps of Engineers, 73, 150, 153, 154, 165; federal jurisdiction of, 160, 164
U.S. Coast and Geodetic Survey, 75
U.S. Department of the Interior, 159
U.S. Environmental Protection Agency (EPA), 8, 160
U.S. Federal Emergency Management Agency (FEMA), 14, 75, 156, 157, 161, 167

VA mortgages, 157
vegetation, 202, 220–221. *See also* beaches, vegetation
velocity zones. *See* V-zones
Virginia Beach, 2, 149, 150–152; development in, 33, 153, 154; hurricanes and, 18, 19; nourishment of, 52, 53; as recreational beach, 44; sand transport on, 115; shoreline retreat in, 43
Virginia Beach City County, 2
Virginia Beach Circuit Court, 153

Virginia Beach Erosion Commission, 152
Virginia Institute of Marine Sciences (VIMS), 75
Virginia Marine Resources Commission, 73, 153
VOR, 23, 24
V-zones, 157

Waddeys, Va., 132
Wallop's Island, 149
Ware Neck, 141
Ware Neck Point, 107, 141
Ware River, 141
Washington, D.C., 3, 15, 17, 77
waste disposal. *See* sewage and waste
Waterford Point, 81
wave(s), 172; erosion by, 26, 28, 50, 65, 96; influence on bulkheads, 64; scour, 59, 63, 172, 175, 189; shadow, 67; storm, 151
weather service, 203. *See also* National Weather Service
weep holes, 59
Weir Point, 118
Westmoreland County, Va., 2, 26, 68
West Point, Va., 2
West River, 125
Wetlands Act, 160; of 1970 (Maryland), 164; of 1972 (Virginia), 164

White and Nelson crab house, 83
White Cliffs, Va., 81
White Road, 88
Wicomico County, Md., 2, 79, 91, 92
Wicomico River, 91
Williams Point, 88
Williamsburg, 2
Willoughby Bay, 144
Willoughby Beach, 151
Willoughby Spit, 53, 115, 144, 145, 147
Wilmer Neck, 110
Wilsonia Neck, 81
Wilton, Va., 137
Windmill Point, 2, 16, 43, 107, 134
winds, 168–169
Wingate, Md., 98
Wise Point, 80
Wolman, M. G., 1
Worten Beach, 90
Worten Point, 90, 113
Wroten Island, 98
Wye Island, 106
Wye River, 104, 106

Yeo Neck, 84
York County, Va., 2; dunes in, 51; erosion rates in, 146; outer shore of, 111; site analysis of, 142
York Point, 111

York River, 2, 141, 142
Yorktown, Va., 17
Yorkville, Va., 142

zoning, 163–164